RESEARCH DATA MANAGEMENT AND DATA LITERACIES

RESEARCH DATA MANAGEMENT AND DATA LITERACIES

Tibor Koltay

CP
CHANDOS PUBLISHING
ELSEVIER An imprint of Elsevier

Chandos Publishing is an imprint of Elsevier
50 Hampshire Street, 5th Floor, Cambridge, MA 02139, United States
The Boulevard, Langford Lane, Kidlington, OX5 1GB, United Kingdom

Library of Congress Cataloging-in-Publication Data
A catalog record for this book is available from the Library of Congress

British Library Cataloguing-in-Publication Data
A catalogue record for this book is available from the British Library

ISBN: 978-0-12-824475-3 (print)

For information on all Chandos publications
visit our website at https://www.elsevier.com/books-and-journals

Publisher: Glyn Jones
Editorial Project Manager: Aleksandra Packowska
Production Project Manager: Vignesh Tamil
Cover Designer: Victoria Pearson Esser

Typeset by STRAIVE, India

Contents

5. New literacies in the age of convergence

6. Roles and education of information and data professionals

Acknowledgments

I express my gratitude to my wife Éva for her patience and support, and to Dr. János Tóth, mathematician, friend, and coauthor of a Hungarian handbook on research and publishing in the natural sciences, for his help in writing this book.

Introduction

Introductions often foretell what will be said in the given work. Sometimes, they also allow an insight into the personal motivations of the author to engage in such an endeavor. This introduction aims to fulfill both functions.

This monograph is in many regards a continuation of an earlier book, titled *Research 2.0 and the Future of Information Literacy*, that we wrote with Sonja Špiranec and László Z. Karvalics (Koltay, Špiranec, & Karvalics, 2016). One of the reasons for mentioning this is that Research 2.0, also called *e-Science* (Hey & Hey, 2006), is still with us, while there are plenty of new developments in academic librarianship.

Some of the fundamental ideas and motivation for writing this book came from two works of Christine Borgman, who directed our attention to the fact that researchers do not need more data, but require the right data. She has also stated that to serve as research data, little data can be just as valuable as big data, or there may be no data at all because relevant data cannot be found, is not available, or does not exist (Borgman, 2007, 2015).

The title of this book is made up of two concepts. The first is *research data management* (RDM) that is the result of the technological developments that made managing big amounts of data possible, triggering thereby the growing interest in research data. The second one appears in the title as *data literacies*, and this requires additional explanation, because it is usually known as *data literacy*. However, the plural term underlines the plurality of approaches and methods in this field. Furthermore, it relates to the endeavor of enabling "the multiplicity and interconnection of data literacy practices with other literacies" (Fotopoulou, 2020). Let us also mention that, in 2014, the author of this book published a paper with a somewhat provoking question in its title: *Big data, big literacies?* (Koltay, 2014). We are certain today that the existence of big data cannot be denied. On the other hand, while it is accepted that there are varied literacies, such a thing as big literacy does not exist, at least not in the literal sense.

This book appears in times when the growing availability of research data has triggered interest in the complex issue of openness, involving the idea of open science (OS), the spread of which is fundamental in academia (although effectuating it seems to be thornier than it was in the case of open access to publications) (Cox & Corrall, 2013). The OS requirements and the management of research data, alongside incentives for data sharing and reuse, also have become high-priority issues, although their acceptance is still varied. Hence the level of preparedness of researchers and academic libraries is not uniform by discipline or by country (Tenopir et al., 2017).

We also see that many (new) literacies have gained ground in our professional lives. There are varied disciplinary approaches to these developments. In this book, most challenges and results related to these developments are portrayed through the lens of library and information science (LIS).

Among the members of this family of literacies, information literacy (IL) has remained a central one. Moreover, in some environments, first of all in academic libraries, IL education is now a practically self-evident task, or close to becoming one, even though there is well-grounded dissatisfaction with its spread and impact. Obviously, we know that a substantial number of new literacies have emerged in the meantime, and among them data literacy has acquired well-deserved importance. This fact has its roots in data literacy's closeness to IL but is also due to recognizing that, without data-literate students and professionals, RDM cannot be successful, or at least may prove ineffective.

The League of European Research Universities (LERU, 2013) has encouraged academic libraries to make efforts to relieve researchers of their technical and administrative burdens. This is undoubtedly a respectable goal, which, however, must be transcended. Therefore libraries should look for extending their collaboration with researchers. More than that, they must make substantial efforts to achieve this goal.

As a rule, from among the many expressions that denote libraries related to research and higher education, such as *university library, college library, research library,* and *academic library,* the latter term will be used in this book. No matter which name is used, services offered by these libraries should be started where appropriate, i.e., when and where their appropriateness and innovativeness are recognized by a number of stakeholders, including communities of researchers, research funders, and research organizations.

The purported target audience of this book includes researchers of all kinds and students in higher education, but the main emphasis is on the duties of various data professionals, first of all academic librarians.

This book includes some amount of practical advice on targeted issues. Nonetheless, the main weight is put on reviewing concepts through a theoretical lens in order to provide conceptual understanding. On the other hand, we deliberately avoid committing ourselves to any single theory.

This book is divided into eight slightly overlapping chapters. Keeping the chapter size uniform was not our ambition, because this would have been an end in itself. We tried instead to adjust the chapter length to the perceived importance of the given content.

The order of the chapters coming after this introduction is as follows:

1. Information, data, text, and document,
2. Information and data ecologies,
3. Data quality, the essential "ingredient,"
4. Research data management,
5. New literacies in the age of convergence,
6. Roles and education of information and data professionals.

The surge of interest in posttruth phenomena and the COVID-19 epidemic inevitably influences literacies, as well as academic libraries and their services; thus these issues are also touched on in this book.

References

Borgman, C. L. (2007). *Scholarship in the digital age: Information, infrastructure, and the Internet*. Cambridge, MA: MIT Press.

Borgman, C. L. (2015). *Big data, little data, no data: Scholarship in the networked world*. Cambridge, MA: MIT Press.

Cox, A. M., & Corrall, S. (2013). Evolving academic library specialties. *Journal of the American Society for Information Science and Technology*, *64*(8), 1526–1542. https://doi.org/10.1002/asi.22847.

Fotopoulou, A. (2020). Conceptualising critical data literacies for civil society organizations: Agency, care, and social responsibility. *Information, Communication & Society*. https://doi.org/10.1080/1369118X.2020.1716041.

Hey, T., & Hey, J. (2006). e-Science and its implications for the library community. *Library Hi Tech*, *24*(4), 515–528. https://doi.org/10.1108/07378830610715383.

Koltay, T. (2014). Big data, big literacies? *Čitalište*, *24*, 3–8.

Koltay, T., Špiranec, S., & Karvalics, L. Z. (2016). *Research 2.0 and the future of information literacy*. Oxford: Chandos Publishing.

LERU. (2013). *LERU roadmap for research data*. Leuven: League of European Research Universities Research Data Working Group. Retrieved from https://www.leru.org/publications/leru-roadmap-for-research-data.

Tenopir, C., Talja, S., Horstmann, W., Late, E., Hughes, D., Pollock, D., … Allard, S. (2017). Research data services in European academic research libraries. *LIBER Quarterly*, *27*(1), 23–44. https://doi.org/10.18352/lq.10180.

CHAPTER

1

Information, data, text, document

Introduction

In 2011 Heidorn stated that data is increasingly recognized as a "first class intellectual object that can undergo quality checks, peer review, distribution and reuse" (Heidorn, 2011, p. 670). A year later Scaramozzino and her colleagues (2012) propounded that, whether produced by small science or big science, "all research data is scientific capital" (Scaramozzino, Ramírez, & McGaughey, 2012, p. 362).

Besides scholarly research, data is often attached to corporate environments. We will come back to this fact in the chapter on data quality. Here let us only mention that, until the early 1990s, business data "was seen as a by-product of running the business and had little value beyond the transaction and the application that processed it; hence, data was not treated as a valuable shared asset" (Al-Ruithe, Benkhelifa, & Hameed, 2019).

As Corrall (2020, p 179) described it, while "data have long been used for decision support and business intelligence, the data revolution has its origins in the fourth scientific paradigm of data-intensive discovery, based on data mining and machine learning with vast amounts of data." Data has "pervaded every sector of the economy and society and is impacting our professional, civic, personal, and social lives."

This has happened at a scale, scope, and complexity never seen before and has been characterized by a range of new technologies, fusing the physical, digital, and biological worlds. With the Internet of Things, the boundaries between the physical, digital, and biological worlds are blurring. On the other hand, there is considerable development related to the amount and availability of freely available, nonproprietary open data, which can be analyzed, visualized, and manipulated.

We are reminded by Buckland (2015, p. 1) that there is no such phenomenon as a *noninformation society*, because all communities, whether human

https://doi.org/10.1016/B978-0-12-824475-3.00003-5

or animal, "are formed by communication, interaction, and collaboration" that depend on information. This is especially true in today's society, which is fueled by information.

We know that data has often been hailed as the *new oil* for science, business, and government, but the question of unjust extraction of value from data is also raised (Loi & Dehaye, 2017). The preeminence of information and data is continually reflected differently, depending on its context and environment. It is therefore important to heed the warning of Bawden and Robinson (2020) that the multiplicity of theories and conceptions about it leads to discontinuities and incompatibilities. Nonetheless, there are varied approaches to information and data.

Today it is recognized that there is an ecosystem that relates not only to information, but also to data. Conceiving of information literacy, data literacy, and several other literacies therefore would not be possible without envisioning processes that relate data to information, or the condition of transforming information to knowledge. There are also complementary issues that concern the nature of text and the characteristics of a document. Most of these issues seem to pertain to the purview of LIS, often addressed as *information science*.

Although we will examine the effect of research data on the work of librarians, let us point out here that this ubiquity of the term *data* and of data itself massively contributes to academic librarians' confusion. As Swanson and Reinhart (2016) explain it, when institutions and/or their libraries call for data services, it is crucial to devise strategic plans that focus on data, although the meaning of data is not always clear when new positions for information professionals are created.

They also add that in traditional reference interviews, qualified librarians knew what they were looking for. The meaning and the goals of data reference interviews are less self-explanatory; therefore it takes much more time to define the term and to tackle the real need. Moreover, among colleagues there may be a lack of a common framework leading to mutual understanding, much less the small nuances of understanding.

Bibliographic data has always been an indispensable part of cataloguing, but other types of quantitative and qualitative data are now central to collection development, research services, and information literacy and data literacy education (Kellam & Thompson, 2016).

Data-intensive work carried out in libraries by data professionals concerns first of all research support. Reference and instruction are directed toward handling and interpreting civic and research, spatial and numeric, and open and restricted data. It is not only academic libraries that are involved, but public libraries are also engaging in data-related activities, especially regarding community and social media data (Bowler, Acker, & Chi, 2019).

Differing approaches to data and information

These snapshots were meant to highlight the premises for identifying the foundational theoretical views related to information and data. Some of these will be explained shortly.

We can start with the three principal uses of the word *information*:

- *Information-as-process*, i.e., the act of informing;
- *Information-as-knowledge*, i.e., knowledge communicated concerning some particular fact, subject, or event;
- *Information-as-thing*, i.e., used attributively for objects, such as data and documents, that are regarded as being informative. This is also intangible, because it cannot be measured directly (Buckland, 1991, p. 351).

Kari (2007) also added to this characterization that information-as-thing should be conceived of as referring to objective packages of cognitive content in a certain form.

Positing information-as-thing reintroduces the concept of a document. On the other hand, it points out the subjective nature of information, by negating the possibility of saying with confidence that it could not be information (Capurro & Hjørland, 2003). As interpreted by Robinson and Bawden (2013), this approach indicates that information is associated with a document.

In the case of information-as-process, information changes a person's knowledge state, while information-as-knowledge is about equating the information with the knowledge that it imparts.

The DIKW approach

The traditional discussion about the nature of data is often based on the DIKW (data-information-knowledge-wisdom) model. This model depicts a hierarchy, where data occupies the bottom of the pyramid (Rowley, 2007).

Frické (2019, p. 135) has argued that the DIKW theory seems to encourage uninspired methodology, because this view suggests that—when collected—existing data is promoted to information and that information answers questions. He adds that this "encourages the mindless and meaningless collection of data in the hope that one day it will ascend to information," and this results in a preemptive acquisition of data.

In his opinion, data is "anything recordable in a semantically and pragmatically sound way" (p. 131). From a semantical point of view, it must be understood as true or false statements. Saying this, Frické also underlines that this view differs from the DIKW account of data, because "in particular circumstances, researchers might regard some recordings as data which report matters that are neither observable nor determinable by instrument" (p. 139).

Although Ma (2013) also rejects the DIKW model by stating that it is "not substantiated by a theory, nor is it supported by empirical evidence." He states nonetheless that, despite this fact, belief in the existence of this hierarchy is widespread.

Depending on the angle and how we look at them, the relationship between data and information can be seen differently.

Data vs information

Viewing information as identical with data is based on the idea that there is no conceptual difference between them, because "both are facts that can be collected, stored, organized, and retrieved." This means that neither of them requires interpretation or meaning. Collecting more data and information enables us to understand the facts and phenomena around us, and thus the progress of science depends on the amount of data and/or information that we collect and analyze (Ma, 2013). From the ontological point of view, data and information are indeed close to each other. However, there is a difference on the epistemological level, where "both data and information assume existence as signs, except that information has meaning attached to the signs." This means that "data in and of itself is not informative. Information is informative because it contains meaning, and document is informative because it carries information." As to their operational characteristics, both are transmittable and retrievable (Yu, 2015, p. 808).

Looking upon information as processed data is also a widely accepted view. One of the sources that laid the foundations of this traditional approach might have been the idea that data and information are conceptually different. Hayes (1969, p. 363) suggested that information results from processed data. Data reflects objective facts that can be collected and stored; however, it becomes information only when processed and organized. Consequently, he postulates that data is not meaningful, because meaningfulness is a characteristic of information. He also assumes that in its raw forms, "data is not understandable and that understanding is not possible without the artificial construction of information infrastructures and that information is the meaningful and understandable form of data."

In 1993, Hayes added that data may be treated as synonymous with "facts." However, "if we want to make a formal distinction between data, as the recorded symbol, and the 'fact' that the data may purport to represent," we can understand that data are not facts (Hayes, 1993, pp. 85–86).

Information may be treated as something objective, quantitative, and mainly associated with data, as is done in physics and technology. However, in the social realm this treatment is rather subjective, qualitative, and often associated with knowledge, meaning, and understanding, while in biology the treatment of information lies between the two (Robinson & Bawden, 2013).

The "process of transforming data into information is a matter of utmost importance. Without the appropriate mechanisms in place, data will stay *raw* and as a result, no information about the physical world and the cultural and social worlds can be generated and the progress of the sciences will be impeded" (Ma, 2013).

Considering information as justifiable claims does not recognize any direct relationship between data and information. Accordingly, it is asserted that information is codetermined in a cultural context. Information may be thus potentially useful in accomplishing a task, or potentially making a person more knowledgeable. Both information itself and the need for information are temporal, because the information found may become knowledge.

As already stated, data is often seen as something that represents the properties of symbols, objects, events, and their environment. In order to be used, data must be in a relevant form, which means that "the nature of difference between data and information is not structural, but functional" (Ackoff, 1989, p. 5), so data does not represent knowledge, but—when processed—it becomes information. While data means not knowing, information answers the question of *What* (Zeleny, 1987), or—as approached by Ackoff (1989)—"information is contained in descriptions, answers to questions that begin with such words as who, what, when and how many. Information systems generate, store, retrieve and process data. Information is inferred from data" (Rowley, 2007, p. 166).

Part of the preceding discussion can be encapsulated by saying that there are researchers who reject the idea of a conceptual fluidity and connections between data and information (Špiranec, Kos, & George, 2019). On the other hand, we see that we can endorse the opinion that it is more helpful to regard information and data as very closely connected phenomena (Bawden & Robinson, 2016).

Today data comprise a set of objects and their features. Data is "not just any arbitrary collection of items existing in some medium such as paper. In a computational environment, *'data' is something a computer can read, transform, and analyse*. This imposes fundamental constraints on how we represent anything" (Manovich, 2019, p. 64, original emphasis). It is not just any arbitrary collection of items, because the medium defines it. The modern understanding of data recognizes data as something that a computer can read, transform, and analyze. The representation of data depends on choices about how we represent anything, i.e., what we choose as objects, what features are chosen, and how these features are encoded. These decisions are important for representing phenomena as data in order to make them computable.

Data interpreted as texts

In the discussion about the nature of data, the digital humanities (DH) occupy a special place and offer a generalizable perspective. As explained by Owens (2011), data is constructed, so it is a species of artifact. In this quality, it is open to interpretation, and thus can hold evidentiary value. As processable information and authored object, created for particular audiences, data can be interpreted as texts, which are open to interpretation and analysis, and as human-made artifacts, datasets have the same characteristics as texts, created for an audience. Owens argues that data "is not in and of itself a kind of evidence but a multifaceted object which can be mobilized as evidence in support of an argument."

The questions that humanists ask about texts are just as relevant to ask of data; thus humanistic approaches to the textual and artifactual qualities of data can offer ways to the interpretation of data. Owens also adds that datasets are, at least indirectly, created by people. In this sense, speaking about raw data may be misleading. Producing datasets is based on decisions about what and how to collect and how the information should be encoded.

We should not forget here that the existence of raw data is contested. Perhaps the best-known argument is that of Gitelman (2013, p. 7), according to whom raw data is "an oxymoron, because data can be good or bad, better or worse, incomplete and insufficient," but does not equal to fact.

Big data

Despite the lack of consensus about the definition and scope of big data, it is usually accepted that it can be characterized by the attributes of Volume, Velocity, Variety, Value, and Veracity. However, there is little understanding of what big data actually is (Janssen, van der Voort, & Wahyudi, 2017).

Big data is not only exhaustive in scope, but can expand in size rapidly. It is also fine-grained in resolution and uniquely indexical in identification; it allows joining different datasets, and adding new fields is easy. The production and use of big data has been facilitated by five technological factors, enumerated by Kitchin (2014). From these, computational power seems to be widely known. Dense internetworking, as well as the ubiquity and pervasiveness of computing, may be coupled to it. Nonetheless, as Ibekwe-SanJuan and Bowker (2017) underline, it is not theoretically satisfactory to define big data by relating it only to the capacity of information technology.

The importance of this argument can be reinforced by the argument of Yu (2015, p. 802), who says that "data in fact refer to the physical entities either waiting to be assigned with meaning (in the case of satellite pictures) or being considered without reference to the meaning already assigned (in the case of lines written by a poet). Although data can be formed preceding to its being assigned with meaning—modern information technologies are producing this type of data in such a scale that they are now called the big data, yet until now, the primary type of data are still those which are composed with meaning."

As Hjørland (2019, p. 685) explains it, "one important aspect about big data is that they are often unintentional traces we leave during all kinds of activities. Their potential to inform somebody about something is therefore less direct compared to data that have been produced intentionally as, for example, scientific databases."

Approaches to big data might be affected by the fact that the digital humanities may offer a type of data that may serve as an alternative to the big data—small data opposition. It is *smart data,* which is defined less by its size than by the fact that its creation involves human agency, i.e., it is actively and purposefully created by people for interpretation. Smart data reduces homogeneity, which can lead to incorrect conclusions. Furthermore, big data is relatively unstructured but massive, while smart data derived from multiple information sources is more structured and relatively small (Schöch, 2013).

The definition of big data should take cultural, technological, and scholarly features into account. In their analysis, Boyd and Crawford (2012) stated that it is not enough to define computing technology that enables gathering, analyzing, linking, and comparing large datasets in order to identify patterns to make economic, social, technical, and legal claims. There is also a mythology that large datasets offer a higher form of intelligence and knowledge never possible before. Let us add that scholarly research may not avoid the latter "side-effect"; thus we need to be wary of it.

As Špiranec et al. (2019) put it, "the *data turn*, the *data deluge*, the *data hype* or *the age of big data* are metaphors, which are used in popular and academic discourses to describe public and institutional awareness of the importance of data. The discussions around data reveal a wide set of perspectives, ranging from expectations and faith in data to emphasizing their risks and pitfalls."

According to Kitchin (2014, p. 1), together with big data, new data analytics is a disruptive innovation, which is "reconfiguring in many instances how research is conducted." On the other extreme, representatives of the enthusiastic and celebratory approaches to the discourse about data emphasize that it carries the power of truth and objectivity, and has the potential to solve the world's most complex and pressing problems (Oliphant, 2017). Those who are more critical, and who share a more nuanced view, question the role, nature, and epistemic qualities of big data by pointing out the contradictory implications and effects of its uncritical use and acceptance for individuals and the society. For instance, Sharon and Zandbergen (2017, pp. 2–3) suggest that instead of "data fetishism," we should adopt a *quantified self*, which is optimized, calculable, and controlled. To achieve this, we should endeavor to "actively engage with data and render it meaningful in and through self-tracking practices," because data can make a "private, subjective and somewhat inaccessible world of feelings and problems more tangible and comparable."

Ekbia et al. (2015) indicate a number of methodological, epistemological, aesthetical, technological, legal and ethical, political, and economic dilemmas, including the problems tied to the relationship between data and our senses and instruments. They raise issues of general understanding and questions of how we can interpret measurements, and how they relate to our knowledge of the world.

Defining the role of big data in scientific research depends on how data in general is defined and identified. Ekbia et al. suggest that, from among the varied views on this complex of issues, we can consider the representational and the relational views, both of which regard data as key empirical evidence.

The representational view "construes data as a reliable representation of reality, produced by the interaction between humans and the world." In contrast, the relational view encourages giving attention to the history of data, explains the critical importance of documenting data management and transformation processes. It also recognizes the expertise of data producers, curators, and the knowledge of those who analyze data.

Emphasis on data as a central component of research can be seen through a theory-centric view of science, positing that "scientific knowledge consists of justified true beliefs about the world." Such approaches favor empirical methods and acknowledge the significance of methods, data, models, instruments, and materials. Datacentric forms of inquiry focus more on the processes through which research is carried out through software and hardware, thus focusing more specifically on computational systems (Leonelli, 2020).

The distinction between relation and correlation is also present in the debates between the proponents of theory-driven science and others who call for an end of theory. Those who want to defend theory may direct attention to the dangers of superficially relying on sheer size. According to a more conciliatory view, it is said that science cannot transcend our human perspective. Debates of a more methodological nature are related to data cleaning that makes the difference between qualitative and quantitative methods needless, while reminding us of the old question of *what counts and what does not* (Ekbia et al., 2015).

Research data

Practically all fields addressed in this book (e.g., research data management, data literacy, and data quality) are influenced by the views on such foundational concepts as data, information, and text. These views are not only important, but their nature is changing as data-driven approaches unfold in our world within and beyond academia.

As indicated in the introduction to this book, a substantial theoretical and practical knowledge has been explained by Christine Borgman. She defined one of the main subjects of this book, research data, as "entities used as evidence of phenomena for the purposes of research or scholarship" (Borgman, 2015, p. 29). This definition is based on the assumption that at some point information becomes data. In the physical and life sciences, most data are usually gathered or produced by observations, experiments, or models. In the social sciences, data come from public records of economic activities, among other places, and researchers may gather or produce it on their own. Data in the humanities may be drawn from records of human culture, including archival materials, published documents, or artifacts.

At the same time, research data is a matter of personal opinion. For the reasons why it can be characterized as personal, we have to look at the ways of defining it in more specific terms. Some data policy documents define data explicitly, but definitions are often based on examples (e.g., facts, observations), or context (e.g., data collected using public funds). The meaning of research data is not only narrower than that of "data," but it should be distinguished from resources such as government statistics or business records (Pasquetto, Randles, & Borgman, 2017). It is also Borgman (2012, p. 1071) who directs our attention to the fact that data and theory are inseparable, as investigations are designed to test or to develop theories, and those theories are used to make sense of the data. Scholarship entails fitting data and theory.

As stated by Tsatsou (2018, p. 1244), "digital technologies generate new phenomena, trends and practices that comprise new objects of research inquiry." This remark makes us aware that in the contemporary environment the importance and use of research data should be spotlighted. It was not recognized sufficiently before, mainly due to the lack of proper technologies. In this sense, we cannot say that research data is a new concept, but thanks to the attention being paid to data-related issues, we experience its rediscovery.

Konkiel (2020) gives examples of research data, by discipline. Biologists work with DNA sequences, microscopy images, morphological data, and images of specimens. Surveys play an important role in sociological researchers' investigations. Beside the responses to survey questions, they also use digitized videos, statistics, or economic indicators. Research in history requires (among others) speech transcripts, digitized photographs and videos, journals, newspaper clippings, and diaries.

Although related not only to research data, gray data must also be mentioned. Gray data is similar to gray literature, which was much more readily distinguishable from scholarly documents in the print world than in today's digital world. Items of gray literature are published in both print and electronic formats today. A decisive criterion of "grayness" is that these publications are controlled by producing bodies whose primary activity is not publishing and thus who are not identical with publishers as we know them (Schöpfel, 2010). Gray literature includes 150 varieties of gray documents, including recorded, referable, and sustainable data, such as databases, datasets, and data papers made available, omitting traditional peer-review processes (Savić, 2019). Nonetheless, gray data is a state-of-the art concept.

Data-driven decision-making

Using data for the development of research and decisions is usually called *data-driven decision-making*. This term is used synonymously with *data-based decision-making*. The expression *data-driven* refers to the internal drive of a given behavior, while *data-based* refers to the decision in itself, therefore lacking the additional motive, i.e., the driving force, which is the key idea of the data-driven decision-making paradigm (Sebestyén, 2019).

Research data can be seen "not just as the result of empirical work or the raw material for analysis, but also as a research object in its own right" (Golub & Hansson, 2017, p. 1100). Nonetheless, it has multiple faces; therefore finding unanimous conceptual understanding of research data seems to be much more complicated today than ever before. However, we should accept that research data consists of "information artefacts that scholars draw upon for evidence in supporting research claims, and producing new knowledge" (Sans & Night, 2019). Raw data is not the only form of evidence, and therefore we can and should rely on reusing research data (Borgman, 2015).

Documents in the light of information and data

When contemplating the similarities and differences between data and information, it is worthwhile to address the concept of the document, which falls into the category of information-as-thing.

As Furner (2016) expressed it, the relation between data and documents is somewhat confused in the literature, first of all due to the lack of consensus about the conceptual relationship between data and document. Nonetheless, the default position appears to be that all documents are in some sense made up of data. The possible reason for this is that "information about the property of a material is incomplete without information about the precision and the conditions of acquiring the given dataset, obtained from particular documents" (Spang-Hanssen, 2001, p. 3).

Documents are available in different media types and on varied physical carriers. They are objects, perceived as signifying something. Moreover, they are concerned with evidence: evidence implies facts, while meaning is constructed in the mind of the observer. The phenomenological status of a document is therefore not inherent, but attributed to an object, whereas its understanding is based on shared cultural codes (Buckland, 2012, p. 3).

Similarly, Otlet (1934) stated that the document captures what we have evaluated and created in our minds. He defined documents as "media inscribed with data; the representation of physical and abstract things," adding that "things themselves may be considered documents when used as such in a study or an argument" (Le Deuff & Perret, 2019, p. 1464).

Nielsen and Hjørland (2014) affirm that data recorded and supplemented with metadata should be considered a type of document. Nonetheless, the characterization of data, information, knowledge, and document by Yu (2015) shows that there is some controversy involved in this question. At the operational level, we see that both data and documents are transmittable and retrievable (Yu, 2015).

However, when looking at the term *document* from a somewhat different angle, we see that it can denote any physical information resource, rather than being limited to specific text-bearing objects. In this sense, a table of numbers can be considered as text, as a document, or as data, while texts that are analyzed statistically could be regarded as data (Buckland, 1991; Latham, 2012).

Information storage and retrieval systems store and retrieve information-as-thing (Buckland, 1991). Translated to data and recorded as documents, data are the only thing that can be managed (Schöpfel, Farace, Prost, Zane, & Hjørland, 2020). Data must be accompanied by other information that can only be found in documents, as no one should request and/or deliver data without information about the precision of the data and about the conditions under which the given dataset(s) were obtained.

If we accept that there is no conceptual difference between data and information, because both can be retrieved, collected, and stored, neither data nor information are subject to interpretation (Ma, 2013).

References

Ackoff, R. L. (1989). From data to wisdom. *Journal of Applied Systems Analysis, 16*(1), 3–9.

Al-Ruithe, M., Benkhelifa, E., & Hameed, K. (2019). A systematic literature review of data governance and cloud data governance. *Personal and Ubiquitous Computing, 23*(5/6), 839–859. https://doi.org/10.1007/s00779-017-1104-3.

Bawden, D., & Robinson, L. (2016). Information and the gaining of understanding. *Journal of Information Science, 42*(3), 294–299. https://doi.org/10.1177/0165551515621691.

Bawden, D., & Robinson, L. (2020). Still minding the gap? Reflecting on transitions between concepts of information in varied domains. *Information, 11*(2), 71. https://doi.org/10.3390/info11020071.

Borgman, Ch. L. (2012). The conundrum of sharing research data. *Journal of the American Society for Information Science and Technology*, *63*(6), 1059–1078. https://doi.org/10.1002/asi.22634.

Borgman, C. L. (2015). *Big data, little data, no data: Scholarship in the networked world*. Cambridge, MA: The MIT Press.

Bowler, L., Acker, A., & Chi, Y. (2019). Perspectives on youth data literacy at the public library: Teen services staff speak out. *Strategic Library*, *64*, 17–25. Retrieved from https://www.libraryspot.net/SL/SL_07.2019/SL_Perspectives.pdf.

Boyd, D., & Crawford, K. (2012). Critical questions for big data: Provocations for a cultural, technological, and scholarly phenomenon. *Information, Communication & Society*, *15*(5), 662–679. https://doi.org/10.1080/1369118X.2012.678878.

Buckland, M. (1991). Information as thing. *Journal of the American Society for Information Science*, *42*(5), 351–360. https://doi.org/10.1002/(SICI)1097-4571(199106)42:5<351::AID-ASI5>3.0.CO;2-3.

Buckland, M. (2012). What kind of science can information science be? *Journal of the American Society for Information Science and Technology*, *63*(1), 1–7. https://doi.org/10.1002/asi.21656.

Buckland, M. (2015). Document theory: An introduction. In *Records, archives and memory*. Zadar: University of Zadar.

Capurro, R., & Hjørland, B. (2003). The concept of information. In B. Cronin (Ed.), *Vol. 37. Annual review of information science and technology* (pp. 343–411). New York, NY: Wiley.

Corrall, S. (2020). Databrarian Ed? Preparing information specialists for participation in an open datafied society. In M. Weaver, & L. Appleton (Eds.), *Bold minds: Library leadership in a time of disruption* (pp. 179–210). London: Facet.

Ekbia, H., Mattioli, M., Kouper, I., Arave, G., Ghazinejad, A., Bowman, T., … Sugimoto, C. R. (2015). Big data, bigger dilemmas: A critical review. *Journal of the Association for Information Science and Technology*, *66*(8), 1523–1545. https://doi.org/10.1002/asi.23294.

Frické, M. (2019). The knowledge pyramid: The DIKW hierarchy. *Knowledge Organization*, *49*(1), 33–46. https://doi.org/10.5771/0943-7444-2019-1-33.

Furner, J. (2016). "Data": The data. In *Information cultures in the digital age* (pp. 287–306). Wiesbaden: Springer.

Gitelman, L. (2013). Introduction. In L. Gitelman (Ed.), *'Raw data' is an oxymoron* (pp. 7–14). Cambridge, MA: MIT Press.

Golub, K., & Hansson, J. (2017). (Big) data in library and digital science: A brief overview of some important problem areas. *Journal of Universal Computer Science*, *23*(1), 1098–1108. https://doi.org/10.3217/jucs-023-11-1098.

Hayes, R. M. (1969). Education in information science. *American Documentation*, *20*(4), 362–365. https://doi.org/10.1002/asi.4630200419.

Hayes, R. M. (1993). Measurement of information and communication: A set of definitions. In J. R. Schement, & B. D. Ruben (Eds.), *Vol. 4. Between communication and information* (pp. 81–104). Piscataway, NJ: Transaction Publishers.

Heidorn, P. B. (2011). The emerging role of libraries in data curation and e-science. *Journal of Library Administration*, *51*(7/8), 662–672. https://doi.org/10.1080/01930826.2011.601269.

Hjørland, B. (2019). Data (with big data and database semantics). *KO Knowledge Organization*, *45*(8), 685–708. https://doi.org/10.5771/0943-7444-2018-8-685.

Ibekwe-SanJuan, F., & Bowker, G. C. (2017). Implications of big data for knowledge organization. *Knowledge Organization*, *44*(3), 187–198. https://hal.archives-ouvertes.fr/hal-01489030.

Janssen, M., van der Voort, H., & Wahyudi, A. (2017). Factors influencing big data decision-making quality. *Journal of Business Research*, *70*, 338–345. https://doi.org/10.1016/j.jbusres.2016.08.007.

Kari, J. (2007). Conceptualizing the personal outcomes of information. *Information Research*, *12*(2). paper 292 Retrieved from http://informationr.net/ir/12-2/paper292.

Kellam, L. M., & Thompson, K. (Eds.). (2016). *Databrarianship: The academic data librarian in theory and practice*. Chicago, IL: Association of College and Research Libraries.

Kitchin, R. (2014). Big data, new epistemologies and paradigm shifts. *Big Data & Society*, *1*(1). https://doi.org/10.1177/2053951714528481.

Konkiel, S. (2020). Assessing the impact and quality of research data using altmetrics and other indicators. *Scholarly Assessment Reports*, *2*(1), 13. https://doi.org/10.29024/sar.13.

Latham, K. F. (2012). Museum object as document: Using Buckland's information concepts to understand museum experiences. *Journal of Documentation*, *68*(1), 45–71. https://doi.org/10.1108/00220411211200329.

Le Deuff, O., & Perret, A. (2019). Hyperdocumentation: Origin and evolution of a concept. *Journal of Documentation*, *75*(6), 1463–1474. https://doi.org/10.1108/JD-03-2019-0053.

Leonelli, S. (2020). Scientific research and big data. In E. N. Zalta (Ed.), *The Stanford encyclopedia of philosophy*. https://plato.stanford.edu/entries/science-big-data/.

Loi, M., & Dehaye, P. O. (2017). If data is the new oil, when is the extraction of value from data unjust? *Filosofia e Questioni Pubbliche*, *7*(2), 137–178.

Ma, L. (2013). Is information still relevant? *Information Research*, *18*(3). paper C33. Retrieved from http://InformationR.net/ir/18-3/colis/paperC33.html.

Manovich, L. (2019). Data. In H. Paul (Ed.), *Critical terms in futures studies in futures studies* (pp. 61–66). Cham: Palgrave Macmillan. https://doi.org/10.1007/978-3-030-28987-4.

Nielsen, H. J., & Hjørland, B. (2014). Curating research data: The potential roles of libraries and information professionals. *Journal of Documentation*, *70*(2), 221–240. https://doi.org/10.1108/JD-03-2013-0034.

Oliphant, T. (2017). A case for critical data studies in library and information studies. *Journal of Critical Library and Information Studies*, *1*(1). https://doi.org/10.24242/jclis.v1i1.22.

Otlet, P. (1934). *Traité de documentation. Le livre sur le livre*. Bruxelles: Palais Mondial.

Owens, T. (2011). Defining data for humanists: Text, artefact, information or evidence. *Journal of Digital Humanities*, *1*(1), 6–8.

Pasquetto, I. V., Randles, B. M., & Borgman, C. L. (2017). On the reuse of scientific data. *Data Science Journal*, *16*(8), 1–9. https://doi.org/10.5334/dsj-2017-008.

Robinson, L., & Bawden, D. (2013). Mind the gap: Transitions between concepts of information in varied domains. In F. Ibekwe-SanJuan, & T. M. Dousa (Eds.), *Theories of information, communication and knowledge* (pp. 121–141). Cham: Springer.

Rowley, J. (2007). The wisdom hierarchy: Representations of the DIKW hierarchy. *Journal of Information Science*, *33*(2), 163–180. https://doi.org/10.1177/0165551506070706.

Sans, S., & Night, W. S. (2019). Introduction. In *HathiTrust: Large-scale data repository in the humanities*. https://akthom.gitbooks.io/hkdcws/content/Introduction/Introduction.html.

Savić, D. (2019). When is 'grey' too 'grey'?—A case of grey data. *Grey Journal (TGJ)*, *15*(2), 19–24.

Scaramozzino, J. M., Ramírez, M. L., & McGaughey, K. J. (2012). A study of faculty data curation behaviors and attitudes at a teaching-centered university. *College & Research Libraries*, *73*(4), 349–365. https://doi.org/10.5860/crl-255.

Schöch, C. (2013). Big? Smart? Clean? Messy? Data in the humanities. *Journal of Digital Humanities*, *2*(3), 2–13. Retrieved from http://journalofdigitalhumanities.org/2-3/big-smart-clean-messy-data-in-the-humanities/.

Schöpfel, J. (2010). Towards a Prague definition of grey literature. In *Twelfth international conference on grey literature: Transparency in grey literature. Grey tech approaches to high tech issues* (pp. 11–26). Retrieved from https://archivesic.ccsd.cnrs.fr/sic_00581570/.

Schöpfel, J., Farace, D., Prost, H., Zane, A., & Hjørland, B. (2020). Data documents. In *ISKO encyclopedia of knowledge organization*. Retrieved form https://www.isko.org/cyclo/data_documents.htm.

Sebestyén, E. (2019). A pedagógiai adatvezérelt döntéshozatal: elméleti megközelítések és vizsgálati lehetőségek. *Magyar Pedagógia*, *119*(4), 287–312. https://doi.org/10.17670/MPed.2019.4.287 (in Hungarian).

Sharon, T., & Zandbergen, D. (2017). From data fetishism to quantifying selves: Self-tracking practices and the other values of data. *New Media & Society, 19*(11), 1695–1709. https://doi.org/10.1177/1461444816636090.

Spang-Hanssen, H. (2001). How to teach about information as related to documentation. *Human IT, 5*(1), 125–143. https://humanit.hb.se/article/view/168.

Špiranec, S., Kos, D., & George, M. (2019). Searching for critical dimensions in data literacy. *Information Research, 24*(4). paper colis1922. Retrieved from http://InformationR.net/ir/24-4/colis/colis1922.html.

Swanson, J., & Reinhart, A. K. (2016). Data in context: Using case studies to generate a common understanding of data in academic libraries. *Journal of Academic Librarianship, 42*(1), 97–101. https://doi.org/10.1016/j.acalib.2015.11.005.

Tsatsou, P. (2018). Literacy and training in digital research: Researchers' views in five social science and humanities disciplines. *New Media & Society, 20*(3), 1240–1259. https://doi.org/10.1177/1461444816688274.

Yu, L. (2015). Back to the fundamentals again. A redefinition of information and associated LIS concepts following a deductive approach. *Journal of Documentation, 71*(4), 795–816. https://doi.org/10.1108/JD-12-2014-0171.

Zeleny, M. (1987). Management support systems: Towards integrated knowledge management. *Human Systems Management, 7*(1), 59–70. https://doi.org/10.3233/HSM-1987-7108.

Further reading

Buckland, M. (2011). Data management as bibliography. *Bulletin of the American Society for Information Science and Technology, 37*(6), 34–37. https://doi.org/10.1002/bult.2011.1720370611.

CHAPTER

2

Information and data ecologies

Introduction

Research data is on its way to becoming part of an information ecosystem that emphasizes reproducibility, sharing, and reuse of data (Pasquetto, Randles, & Borgman, 2017). This statement confirms that there is no strict line of division between information- and data-related phenomena. In this spirit, we can say that the goal of information ecology is to examine the complex relationships between human beings in information places and spaces, manifested in the form of sense-making (Steinerová, 2010). Although originally conceived to depict ecological dimensions of information literacy, the factors related to open science, expertise, and methodologies appearing in Fig. 1 are appropriate to general information and data ecologies as well.

These and similar approaches are broad enough to encompass the concepts of open data and gatekeeping and the efforts to foster data sharing and reuse, and to explain the phenomenon of information (and data) overload and posttruth phenomena, to mention the most important ones. As was declared earlier in this book, some of these issues are of interest mainly for researchers, but we must also think about students, first of all graduate and doctoral ones. Obviously, all problems deserve the attention of academic librarians.

Open science and open data

As proclaimed by the League of European Research Universities, open science requires a "culture change in the way stakeholders in the research, education, and knowledge exchange communities create, store and deliver the outputs of their activity" (LERU, 2018, p. 3).

In the United States, the National Science Foundation (NSF) in 2011 took an important step toward the openness of data by requiring the preparation of data management plans (DMPs) by their grantees. Broader

Research Data Management and Data Literacies
https://doi.org/10.1016/B978-0-12-824475-3.00001-1

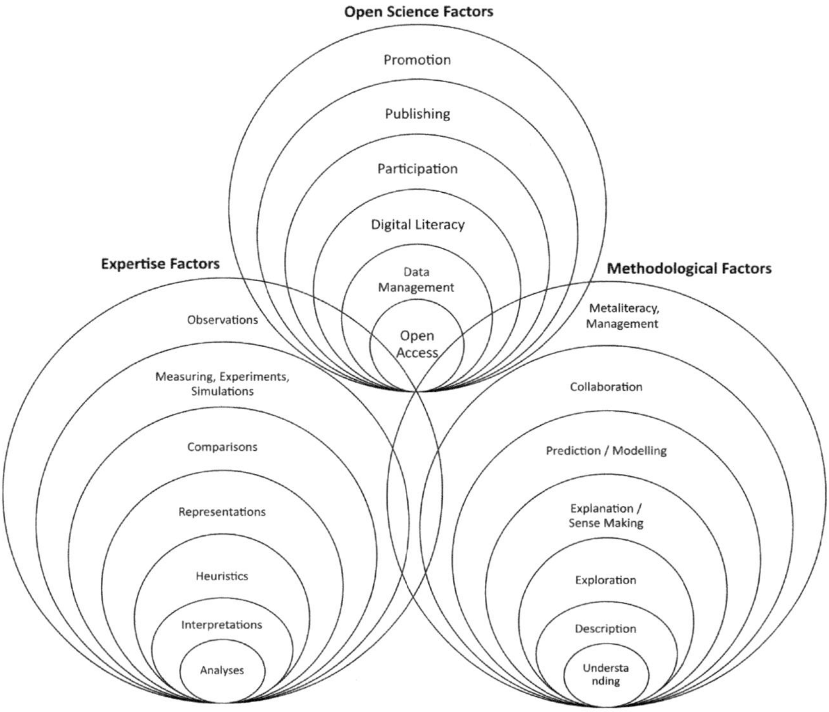

FIG. 1 Ecological framework of research information interactions. From Steinerová, J. (2017). Modelling information environment in ecological contexts: A qualitative study of information behaviour of researchers in Slovakia. In *Knižničná a Informačná Veda (Library and Information Science) XXVII* (pp. 11–27). Bratislava: Univerzita Komenského v Bratislave.

open science and open data initiatives appeared with the adoption of the Fair Access to Science and Technology Research Act in 2013 (US Congress, 2013). This culture change is also strongly supported by the European Commission through the European Open Science Cloud (EOSC), which aims at creating a trusted, virtual, federated environment for hosting and processing research data across borders and scientific disciplines in the European Union. Central elements of EOSC are the European Data Strategy and the European Research Area. The EOSC is now developed under the Horizon Europe 2021–27 program (EC, n.d.).

The openness of data is perhaps one of the most problematic concepts, full of controversies and competing narratives. This is especially pronounced when we see that it requires the fewest restrictions and the lowest possible costs, while there are numerous conditions for open data that are difficult to comply with (Pasquetto et al., 2017). One of the reasons for this is the recursive nature of data, allowing its reuse in a variety of ways. The life of data extends beyond the point of generation and capture, being determined by the composition of the dataset and the algorithms involved in data processing (Pangrazio & Sefton-Green, 2020).

As is well known, the main targets of open access (OA) are journal articles. However, open data is not only different, but must also face a larger number of complex obstacles than those that OA had to overcome. Consequently, the worldwide challenge to give appropriate attention to research data is far from being properly and unanimously recognized, or evenly spread. Nonetheless, when acknowledged, this challenge is often answered by academic libraries in many countries, who provide a wide array of services that make research data management (RDM) possible and efficient; this is addressed in Chapter 4, Research Data Management.

When planning the introduction or expansion of services based on open data, it seems to be worthwhile to accept the position of Van Tuyl and Whitmire (2016, pp. 1–2). In their opinion, "the bar should be set relatively low for the usability of data in order to facilitate the downstream goals of sharing such as validation of results, reproducibility of research, and use of data for novel science."

Openness of data is then related to answering questions about the ownership of data. When the question "Who owns your data?" is asked, the answer may be relatively clear if research is funded, but this category does not cover all situations, as there may be unfunded research, or the data in question may come from a pilot project. As described by Zilinski, Gobens, and Briney (2015), a number of questions arise in this regard. They may cover situations when a researcher leaves the institution. Another question is about the length of keeping datasets and the ways of discarding them. There may be doubts surrounding the manner of setting up data policies for a given researcher, team, or lab. Deciding about the ownership between collaborating teams also may prove deeply complicated. Obviously, several problems can surface, but academic libraries should be ready to provide guidance. A number of services designed to help in these situations are part of research data management that is described in more detail in Chapter 4.

Openness is closely tied to data sharing, the reasons for which are manifold. For instance, reusing data can save researchers' time, as it prevents "reinventing" results already attained by others. Providing services for data sharing promises opening new prospects for academic libraries, but it is not always easy to convince researchers of the necessity of accepting data sharing.

Neither is it a simple task to persuade librarians to introduce related new services. The problem is that working around open data is a challenge for librarians, because many of them do not possess analytical skills and lack the abilities needed for working confidently in this changing information environment (Kirkwood, 2016). We will deal with some of the problems of meeting this challenge in Chapter 4.

In this regard, we should keep an eye on the mass of useful skills that have been acquired by librarians in the traditional, textual information environment that, in many cases, can be transferred to finding, collecting, and presenting data. Obviously, in the digital environment, these abilities should be updated and mobilized in collaboration with researchers. Moreover, there is a need for changing librarians' appreciation of technical and socioethical issues related to research data (Robinson & Bawden, 2017).

Librarians should work toward transcending their present status of being auxiliaries and bearing researchers' technical and administrative burdens. They could attempt to achieve true collaboration with researchers. Obviously, such moves would require focused and intensive work by library managers, who must understand the advantages of RDM. It is crucial, therefore, that these managers choose appropriate services and eliminate inappropriate organizational structures and inflexible job descriptions. Last but not least, they must acquire backing from their whole staff. Many of the efforts to build data-based services, especially if they are done in a cooperative way, can yield a branding effect, similar to the appreciation that libraries have enjoyed by being trustworthy institutions and sources of information (Burton & Lyon, 2017).

The analysis of academic librarians' RDM experiences in 2018 showed concerns about not being considered as RDM support providers, due to libraries' late entry into this field. This may have been caused by the fact that researchers started supporting activities themselves, because the library did not show any clear intention to provide services. Nonetheless, whatever earlier perceptions researchers had or still might have, librarians want to be engaged in supporting the research communities. Moreover, they intend to educate researchers in fields where it can be supposed that the library profession has accumulated expertise and could be expected to change perceptions about librarians and the value of libraries (Faniel & Connaway, 2018). Unfortunately, librarians often overrate the value of their services, especially if their approach is not holistic, i.e., they see the library only, without giving enough attention to the whole institution (Klain Gabbay & Shoham, 2019).

When identifying the trends and issues affecting academic libraries in, 2020, the members of the Research Planning and Review Committee that carried out a biennial environmental scan in the member libraries of the Association of College and Research Libraries (ACRL) found that open data is facing obstacles, especially in health sciences and social sciences research. The attention paid to ethical issues of the scientific practice and its technical workflow was found to grow. This move involved open access, open data, and open science. The Committee also identified increased attention toward FAIR (Findability, Accessibility, Interoperability, and Reuse) principles, which nonetheless remained relatively unknown.

They also stated that, despite positive developments observed in 2019, researchers seem to be slow to respond to open data mandates, because sharing research data brings with it many challenges for them (Benedetti et al., 2020).

At the same time, we should not lose sight of the fact that data sharing and reuse are not restricted to research data and academic libraries, but there is interest in disseminating the idea of open data and data sharing in the sphere of public libraries (Copeland, Yoon, & Zhang, 2020).

The existence and activities of the Research Data Alliance (RDA) are undoubtedly worth mentioning here. This community-driven initiative was launched in 2013 by the European Commission, the US National Science Foundation and National Institute of Standards and Technology, as well as the Australian Government's Department of Innovation "with the goal of building the social and technical infrastructure to enable open sharing and re-use of data."

As of March 2021, this organization had more than 11,000 individual members from 145 countries. These members are engaged in 37 Working Groups and 56 Interest Groups. There are also 50 organizational and 12 affiliated members.

RDA provides "a neutral space where its members can come together through focused global Working and Interest Groups to develop and adopt infrastructure that promotes data-sharing and data-driven research, and accelerate the growth of a cohesive data community that integrates contributors across domain, research, national, geographical and generational boundaries" (RDA, n.d.).

Despite notable developments, "difficulties arise because researchers are either unaware of the benefits and evolving landscape of open science, do not possess the skills and resources to manage their own data effectively, or do not see its value because it is not appropriately rewarded by the current academic system" (Savage & Cadwallader, 2019, p. 1). In this situation, a new type of research skill for gathering information and creating new knowledge is required in situations where researchers need not only to choose the appropriate methodology but also to decide how to manage and promote their outputs. To help researchers in supporting associated good practices of research data management is a challenging task for academic librarians, who must develop skills related to scholarly communication themselves. The Research Support Ambassador Programme at the University of Cambridge (United Kingdom), for instance, was directed toward training library staff who wish to acquire in-depth knowledge of research support (Sewell & Kingsley, 2017). The Ambassador Programme has been supplemented by the Data Champions Programme, which recruited volunteers from across the University in order to provide a sustainable and community-driven initiative that enables advising researchers, who are members of their own research communities. Giving proper advice on handling of research data and promoting good RDM could not have been possible without being heavily facilitated by a central unit, i.e., the Office of Scholarly Communication, Cambridge University Library. More details on this latter program can be seen at https://www.data.cam.ac.uk/.

Data sharing and reuse

Not surprisingly, acquiring the "right" data is vital in any research area, because data that is not appropriate for the purpose, or that is of low quality, may distort results and make research unreliable. Consequently, determining what data is valuable and amenable for being cited partly depends on the quantity and quality of the data that is released and made publicly available. However, it is also difficult to know which datasets researchers choose to share; neither do we clearly see when and why they reuse data. Moreover, data may serve as a subject of exchange for further data, or be an invitation to collaborate. Although not many studies have formally defined the term *reuse*, it is generally understood as the use of data by someone who did not collect it. Therefore reuse refers to a secondary use of data that are not defined by their original purpose but are intended to address new problems.

As explained by Poole (2016, p. 964):

> "Data sharing allows scholars to provide descriptive or historical context, to detect fraud or to address disputes, to reproduce or to verify research findings, to break new ground in research methodology, to make findings generated by publicly funded research available, to enable meta-analysis, to increase citation, to reduce loss, to enrich pedagogy, to generate new knowledge from familiar data and thus to leverage investments, to avoid the costs of re-collecting data, and to foster economic development."

Although influencing different disciplines to a varied extent, the traditional behavior of the majority of researchers often prevents the transparent provision and reuse of research data.

As described by Oßwald (2021, pp. 277–279), in order to change this situation, all researchers are challenged. Such challenges can be met by "establishing infrastructures to support researchers through financial and personal incentives, as well as reward mechanisms for the preparation and free accessibility of research data. There are obviously barriers to such developments, because the motivation to find solutions to research questions is coupled with the need for gaining scientific reputation with these solutions. Accordingly, in most areas, the science system is geared toward individuals and the results that they have achieved and possibly published for the first time. Science-related rewards such as titles, honors, positions and, in particular, research grants are usually awarded for the results achieved. In such a system, which is strongly geared toward personal performance and career, research data, which was collected with great effort and in lengthy processes, is capital, which is only made transparent if it has already been comprehensively evaluated and transformed into publications, or if there are other necessities to make them open."

If looking at the possible practical side of sharing, we can see that it has varied forms, including private exchanges between researchers, posting datasets on websites, and depositing datasets in archives and repositories. Attaching data as supplemental material in journal articles is also a possibility, and—as described shortly—providing data via journal papers, as well as data papers, determined mainly to be published in data journals, devised expressly to foster data sharing. These alternatives are apparently gaining ground today.

Data-sharing policies vary widely by research domain, country, and agency. At the same time, they also show many similarities. The arguments for sharing data include reproducing research, and leveraging investments in research. However, data sharing "in its simplest form, is nothing more than merely releasing or posting data. The real benefit will be achieved only if the available data is used, or reused by others."

The ability to share and reuse depends on decisions made promptly about how to describe and manage data. There is also a need for documenting data in ways that make the data reusable early and enable representing them as citable objects. Data citation mechanisms can support these functions, although the presence of individual citations, per se, may not be enough to enable the interpretation of data or the documenting of its provenance (Pasquetto et al., 2017, p. 1).

The commitment to making knowledge more easily accessible to others arguably goes back to the advent of scientific journals in the mid-17th century. However, the best-known example of modern views relating to open science was the Human Genome Project, which was one of the largest publicly funded, international collaborative ventures in the 1990s (Majumder & McGuire, 2020).

Today, it is a requirement for many articles to be accompanied by the underlying research data. As Gold et al. (2019, p. 4) explain it, such policies support data sharing in order to "reduce transaction costs, promote data re-use, increase rigor and reproducibility, decrease redundant research, better involve patients, consumers, and others, facilitate researcher transparency in sharing processes and results, and improve connections with a larger variety of actors to produce more innovative approaches and solutions over the medium to long terms."

In general, it is clear that reuse must be enabled by the provision of high-quality, curated data resources (Neylon et al., 2019). As data sharing is a multifaceted endeavor, campaigning for it may achieve the intended goals, but this does not happen without some frustrations. As Chawinga and Zinn (2020, p. 109) put it, "the concept of data sharing is complex as it operates amidst an interplay of factors on a continuum where some factors create a flourishing data sharing environment while others act in opposition."

On the other hand, the words of Smith (2014, p. 56) have remained valid. He stated that "primary research data is not yet well understood as a research product or as intellectual property with defined roles and responsibilities for its production, publication, reuse, and preservation."

The need for data sharing and reuse dictates the use of research data life-cycle models (Federer, 2016), which appear in Chapter 4. Our present argument also concerns attributes of data quality, discussed in more detail in an entire chapter of this book dedicated to quality issues.

The benefits of reuse are recognized by some researchers, although this is not yet the norm in every discipline. In the absence of standards effective for ensuring data quality, the pursuit of trustworthiness and the ability to study new problems becomes an important motivating factor for reuse. To be successful, reuse requires data curation that provides the means for assuring data integrity and authenticity, and enables access to high-quality data that can be trusted. Hence, trust (trustworthiness) is an important driver of data sharing and reuse, whereby important venues of sharing and reusing are data repositories, as they act as trusted intermediaries between those who give data and those who receive it (Yoon & Lee, 2019).

Obviously, there are a number of other tools and indicators beyond trust that verify the integrity of digital objects. This role may be played by unique and persistent identifiers, stable and persistent links between digital objects and digital signatures. Nonetheless, the eventual presence of sophisticated technology does not supersede trust, based in the individuals and the social institutions, eminently libraries, involved (Blanchette, 2012).

The reuse of data is driven more by the social norms in disciplines or research than by personal motivation. This also means that interests in "nurturing a culture of data reuse would take time, especially if a discipline or community has a tradition of not allowing data reuse" (Yoon & Kim, 2017, p. 231). It is therefore not surprising that, due to the "lack of knowledge about or access to existing data, many researchers create new primary data rather than reuse existing data" (Bugaje & Chowdhury, 2018, p. 340).

Data sharing and reuse are also influenced by the way digital data moves between social actors. Related to these issues, Bates (2018) directs our attention to the phenomenon of data friction that can also be physical, but is mostly induced by social environments. Data friction is a barrier to scientific advancement and can occur at any stage of the data life cycle, mainly impeding the movement and sharing of data in some form.

Nafus (2014) observed that the source of friction could be the quantity of data, but there are institutional and political factors as well. These factors also involve the costs in time, energy, and attention required by collecting, checking, storing, moving, receiving, and accessing data (Edwards, Mayernik, Batcheller, Bowker, & Borgman, 2011). In addition to this, scientific datasets may be unreadable, corrupted, erased, or otherwise impossible to recover. There are hurdles such as size, reliance on software, distributed ownership, dispersed storage, or lack of documented provenance. An important bar set against sharing and reusing is in the thinking and practices of scholars "who regard their data as having little value beyond the confines of a small group, a specific project, or a specified period" (Ogburn, 2010, p. 242). If shared, many aspects of data may become public good, as datasets document the research problem addressed and describe how data was collected. They show what analyses were performed on the given dataset, and how the results were interpreted.

While data sharing is a desired outcome of making results of publicly funded research available to the public, we go a step further when reusing already existing data, because this enables asking new questions, using data from researchers who were not the original investigators. The act of enabling reuse promises more for prospective users than it can benefit the producers of data. It also allows reproducing or verifying research, thus advancing the state of scholarship and innovation (Borgman, 2012).

From among the impediments, we find that preparing truly reusable data requires an investment in time and resources as well as skills, many of which are lacking. When deciding about sharing and reusing a dataset, there are still hurdles before the researcher that can be overstepped by making decisions based on answering the following eight questions:

1. Does a suitable dataset exist?
2. Where is it located?
3. Is the copy too deteriorated and/or obsolete?
4. Is its use permitted?
5. Is it standardized (interoperable) enough to be usable with acceptable effort?
6. Is what the given dataset represents clearly described?
7. Are the lineage, version, and error rate understood, trusted, and acceptable?
8. Is it appropriate for the intended purpose?

In practice, these questions may not be answered with a simple yes or no, but form a chain: "If you learn that a dataset exists you may not be able to locate a copy; if you can locate a copy it may not be usable; if it is usable, you may not be able to obtain permission; and so on. Any problem might prevent re-use, but, if resolved, another might still prevent it" (Buckland, 2011, p. 35).

There are several other areas of continual concern, which should be taken into account if someone wants to make data widely available in order to advance research. For instance, it is important "to protect the professional interests of data generators through abiding by fair publication policies and clear procedures in order to ensure that researchers, who spend the time and intellectual capital on generating data, can get credit for their work" (Majumder & McGuire, 2020, pp. 168–169).

As Faniel and Jacobsen (2010) see it, those researchers who work in the scientific and technical computing communities argue for data sharing, and are well aware of the difficulties involved in reusing others' data, because assessing the veracity and integrity of a given dataset requires domain expertise, and that assessment depends upon the extent of the documentation available.

It is also noteworthy to know that raw data is not the only form of evidence; thus we can and should build on reusing research data (Borgman, 2015). Reuse is influenced by researchers' perceptions, personal knowledge, skills, and understanding of errors, among other factors. These factors also determine how reusers judge data quality (Borgman, 2012). The relevance of authority, answering questions about the creator and the financial supporters of the dataset, as well as about bias and currency of maintenance, are also playing a major role. Accordingly, it is also important to answer several questions, of which the following three are especially important:

- "Who is in charge of checking for quality?
- What process do they use?
- How is missing data handled?" (Zilinski & Sapp Nelson, 2014, p. 2).

Researchers who are experienced in data reuse may have a certain level of confidence in data reuse and face more concerns about the misuse of data. Their intentions to reuse data at their own organizations can positively influence what they choose (Kim & Yoon, 2017).

Reuse is also a foundational factor for determining data quality when standards for ensuring quality are absent, because they may rely on others' expertise in assessing data. If the original data producers are reputable, it can be supposed that they created reputable data, which is based on peer-reviewed publications, displaying transparent methodologies that are consistent with disciplinary norms.

Beyond trustworthiness, accessibility, data credibility, and the quality of documentation, completeness and the ease of operation also contribute to decisions about what data to reuse. Similarly important are naming conventions and units used. An effective tool for identifying reusable data is examining dates of creation and update, and so is data citation (Faniel, Kriesberg, & Yakel, 2016).

The relationship between reuse and quality is not a one-way street, because researchers tend to reuse the datasets produced by researchers.

Quality issues often need to be solved on a case-by-case basis (Mons, 2018). Decisions largely depend on whom they trust, apparently because "many services do not have mechanisms that could ensure the proper quality of datasets uploaded by their users" (Bugaje & Chowdhury, 2018, p. 341).

Besides reuse, we can also consider repurposing. While data reuse can be defined as "using the data more than once for the same purpose, repurposing tends to mean the use of data for a completely different purpose, compared to the original one" (Yoon & Lee, 2019, p. 1246).

Unfortunately, there is an absolute impediment to data sharing and reuse. It is the "no data" phenomenon, i.e., lack of availability of data for scholarly inquiry in some disciplines. This may be caused by the absence of open access to data for many reasons, including issues of privacy and security, or reasons related to competition, as well as data being unusable due to poor documentation or lack of metadata (Borgman, 2015).

Misuse and misinterpretation of data

In the light of the preceding argument, misuse and misinterpretation of data are significant concerns among researchers when sharing their data. The risk of misinterpretation depends on the origin of data, the required documentation, and the effort required on the part of the reuser (Borgman, 2012).

Misinterpretation is not only a practical issue, but a theoretical concern as well. When examining semiotic dimensions of data quality, Huang (2018) has identified three communication gaps that occur during sign transference within different semiotic levels in the DIKW (Data-Information-Knowledge-Wisdom) hierarchy, also mentioned in Chapter 1. A receptivity gap occurs due to the difficulties in access through physical channels. A usefulness gap appears when representation is improper within a particular context. Insufficient meaning of data leads to misinterpretation of data. This phenomenon, called the *interpretation gap*, leads to poor predicates for the process of decision-making that is essential for sharing and reuse.

An earlier series of surveys that examined, among other aspects, data sharing in six European countries showed that the highest level of worries about misuse was found in Slovenia, with 55% of the respondents. In Croatia 49% of professionals felt this. This perception characterized 33% of researchers in the Czech Republic, 25% in Bulgaria, 24% in Iceland, and 19% in Hungary. The share of responses for misinterpretation was 50% in Croatia (Špiranec & Kos, 2017), 49% in Slovenia (Vilar & Zabukovec, 2019), 37% in the Czech Republic (Drobíková, Jarolímková, & Souček, 2017), 25% in Hungary (Jávorszky, Murányi, & Koltay, 2019), 22% in Iceland (Pálsdóttir, 2017), and 6.5% in Bulgaria (Todorova, Garvanova, Peteva, & Avramova, 2019).

The results of a survey at the University of Kentucky (Lexington, KY) also showed that misinterpretation is one of the reasons behind researchers' unwillingness to consider data sharing. However, there were other frequently mentioned problems:

- The confidential, proprietary, or classified nature of information,
- Lack of expertise in data sharing,
- Time or effort required to make data available,
- Intellectual property concerns,
- Lack of appropriate tools for sharing or publishing data,
- Concerns about losing a research advantage (Joo & Peters, 2020, p. 643).

In their metasynthesis of 45 studies and 3 companion reports published between 2003 and 2018 and mostly conducted in North America (60%) and Europe (17%), Perrier, Blondal, and MacDonald (2020) found lack of responsible conduct of research. Neglecting the professional standards, ethical principles, and tacit norms of the given discipline—although relatively independent from reuse—causes inappropriate use of data that may include misuse and misinterpretation. Data integrity that reflects researcher's expectations about the prospects of reusing data available in repositories was not forgotten either.

It was found that researchers feel a lack of incentives and an absence of direct benefits for sharing their data with regard to academic appointment, promotion, recognition, and rewards, because deficiencies in infrastructure may hamper sharing data effectively. It was found that this is particularly true for the case of sensitive data, and the authors had the experience that "the availability of data is marked by researchers' decision making around what they determine is worth sharing data." (p. 16).

Perrier, Blondal, and MacDonald revealed that the lack of incentives for sharing is not the only impediment. Researchers' lack of time, resources, and skills to share their data effectively in public repositories is also a problem. The respondents were aware of the fact that the worth of sharing a given dataset usually depends on subjective decisions.

In this context, it is perhaps not surprising that there are always researchers who would not trust the quality of a dataset not collected by themselves, although the perceived quality of datasets is a source of headaches, because such quality varies depending on the original researchers' and the reusers' intentions, or the purpose of the collected data. Although it may seem self-explanatory, the respondents expected a significant amount of detail, contextual information, and metadata to be included in the documentation accompanying the datasets.

Reproducibility

Reproducibility refers to "measures that would support the ability to verify, reproduce, or potentially replicate the results of research." Services related to data curation and sharing offered by academic libraries provide immediate support in this field.

Recently, evidence from a number of disciplines has emerged that many scientific studies (e.g., in psychology, biology, biomedicine, neuroscience, drug development, chemistry, climate science, economics, and education) may not be reproducible. This has led to declarations that science is experiencing a "reproducibility crisis" and that this crisis reduces the extent to which researchers and the public can trust research, because the related terms are often misused, which leads to ambiguous and inconsistent use (Sayre & Riegelman, 2018, p. 267).

As an example, we can take computational and experimental mathematics. From the list defining the criteria of reproducibility in these fields, we mention the following criteria requirements, which seem to be valid for a broader set of sciences:

- Reviewable, i.e., independently assessed and judged credible, especially by peer review;
- Replicable, i.e., made available for others to duplicate the results (in some cases, this might be restricted to referees or only upon request);
- Open or reproducible, i.e., auditable by being made openly available, comprising well-documented and fully open code and data (Stodden, Bailey, Borwein, LeVeque, & Stein, 2013, p. 10).

As was put down in the *UCL Statement on Transparency in Research*, reproducibility is a "concept which is particularly relevant to the experimental sciences with a quantitative focus. In the Arts, Humanities and Social Sciences it may be more useful to refer to transparency or academic rigour in the use of research methods through the whole research process."

The reproducibility of research methods is "required for research to be replicated. This, in turn, is essential in research contexts where findings must be robust and reproducible in order to form a solid foundation on which to build further knowledge."

In research contexts where reproducibility is possible and appropriate, researchers are strongly encouraged to use measures that support it. These include (but are not limited to):

- Preregistration of study procedures with analysis plans, and use of registered reports where appropriate;
- Transparent reporting of research in line with recognized community guidelines;
- Disclosure of all tested conditions, analyzed measures, and results;

- Transparency around statistical methods (including sample size planning and statistical assumptions and pitfalls);
- Use of preprints;
- Carrying out replication studies;
- Publishing "null" findings (Ayris, 2021, pp. 42–43).

Going FAIR

A number of contemporary requirements set against research data and addressed here fit into the framework of the increasingly recognized and accepted FAIR principles. This acronym covers Findability, Accessibility, Interoperability, and Reuse of digital data, produced either by machines or by humans. Alongside of good data management and data stewardship, applying FAIR principles can foster reuse of published data (Wilkinson et al., 2016). Although sharing and reusing data and a number of related issues have already been discussed, this initiative merits special attention.

With the changes in the types and variety of data, there are apparent bottlenecks in the scholarly communication cycle, because the development of data-driven research provides not only opportunities for researchers, but it is at the same time a major obstacle. In this regard, a good example of overcoming the existing hurdles is by aiming to "accelerate the discovery of new knowledge in the cycle of scholarly communication by making research resources easy to find through a semantically enabled search application." Resource sharing namely requires documentation and organization, which are rarely found in the practice of research laboratories, but dedicated curation teams can be useful, especially if they operate taking account of "the modern library's new mandate to facilitate and accelerate the discovery of new knowledge" (Haendel, Vasilevsky, & Wirz, 2012, p. 5).

The previously mentioned situation of researchers' lack of involvement is reflected among others in the quantitative study of Pálsdóttir (2021), who surveyed 792 academic researchers and doctoral students and received 139 answers (18%). Her main findings show that their familiarity with the procedures of data management is limited, and thus managing data is not a normal practice in these researchers' professional life.

The 2020 *State of Open Data* (*Histories and Horizons*) report based on the answers of over 4500 respondents worldwide shows that the proportion of those researchers who never heard of the FAIR principles changed from 60% to 39%. At the same time, the overall familiarity increased from 15% to 24% (Hyndman & Goodey, 2020).

At present, the implementation and adoption of the FAIR data principles are still in development, and the EOSC initiative, mentioned earlier, is in a work-in-progress stage. The *Turning FAIR into Reality (TFiR) Report* published in 2018 recommended the use of existing community-endorsed certification frameworks for Trustworthy Digital Repositories, in particular CoreTrustSeal, mentioned in the chapter on data quality. Such certification enables securing a baseline level of FAIR-ness to the datasets, and this level may be maintained or even increased over time (Koers et al., 2020).

The environmental scan by the ACRL Research Planning and Review Committee, mentioned earlier, showed that the reasons for the relatively slow acceptance of the open data idea, including adherence to FAIR principles by researchers, were primarily due to feeling suspicious and having apprehensions about the misuse of openly shared data (Benedetti et al., 2020).

Data ethics

There are ethical challenges posed by managing research data. This is not the only problem tied to data science. Nonetheless, the discussion will be restricted here to ethical questions, while further issues will be described in detail in the chapter reflecting on the roles and education of information and data professionals.

By its nature, the shift from information ethics to data ethics has been both semantic and conceptual, but the former was more outspoken. Nonetheless, this topic not only displays interest in moral dimensions and problems, but examines related practices and solutions. Data ethics refines the approaches known from computer and information ethics. It addresses the true invariant of our concerns, and highlights the need for concentrating on the content and nature of computational operations.

According to Floridi and Taddeo (2016, p. 3), data ethics "can be defined as the branch of ethics that studies and evaluates moral problems related to data (including generation, recording, curation, processing, dissemination, sharing and use), algorithms (including artificial intelligence, artificial agents, machine learning and robots) and corresponding practices (including responsible innovation, programming, hacking and professional codes), in order to formulate and support morally good solutions (e.g., right conducts or right values). This means that the ethical challenges posed by data science can be mapped within the conceptual space delineated by three axes of research: the ethics of data, the ethics of algorithms, and the ethics of practices."

Data papers and data journals

Data papers, providing descriptions of methods for collecting, processing, and verifying data are arguably not new, but they are becoming more widespread. This helps in improving the control of data provenance and giving credit to data producers (Pasquetto et al., 2017). These searchable articles are in many respects consistent with standard scientific communication practices, as they describe online-accessible datasets or sets of datasets in a form similar to journal articles (Candela, Castelli, Manghi, & Tani, 2015).

Data papers describe particular datasets or collections of datasets and give background to the data and facts about their physical properties and precision. They are not only citable, but provide crosslinks to data repositories. Moreover, data papers provide descriptions of published research datasets, accompanied by contextual information about their production, acquisition, and potential utility (Schöpfel, Farace, Prost, Zane, & Hjørland, 2020).

Data papers have the potential to receive a high number of citations because they are published in peer-reviewed, open-access data journals. These journals provide contextual information about dataset production and offer appropriate mechanisms and forums for data sharing and reuse (Jiao & Darch, 2020). These journals contain metadata on the acquisition, methods, and processing of specific datasets. The published papers are cross-linked with approved repositories (Hrynaszkiewicz, 2019).

In data journals, researchers do not refer to the datasets themselves, but to the articles that appear in the journals, and they describe online-accessible datasets in a form similar to journal articles. Hence, the existence of data papers reconciles data citation with the traditional reference strategies, because instead of the datasets, the articles are cited. However, data journals do not solve all the data citation challenges (Candela et al., 2015).

As summarized by Schöpfel et al. (2020), perhaps the most relevant aspect is "the integration of data journals and data papers in a complex structure of open access journal platforms and data repositories, academic communities, research projects, conferences, etc. Interconnectedness requires interoperability between platforms and infrastructures but is more than technology, formats and standards, insofar as it means new ways of doing science, including research management, research environment, workflows, etc."

They also state that—approximately from 2008—data journals have been published by big and small academic publishers, learned societies, and university presses. Some of them are subscription-based ones, but the majority use the Gold Open Access business model. To guarantee the utility and reusability of the papers' quality, as well as to assert some of the datasets, most data journals have some kind of peer review in force.

A content analysis of 15 data paper templates from 24 data journals, indexed by the Web of Science by Kim (2020), implies that providing a more standardized set of data paper components would be beneficial for presenting relevant contextual information in a consistent manner.

Openness through citing research data and using alternative indicators

The growing support of the research community for open access and open research, including data sharing and citability, has positively influenced the views on evaluating research data (Piwowar et al., 2018).

This change has affected citations of research data. It is well known that "citations are the cornerstone of knowledge propagation and the primary means of assessing the quality of research." However, citations can serve data attribution in the sense of giving credit to the people involved in the creation and curation of data, as a recognition of their accomplishment. It also indicates data connection, because data citation methods are required to connect scientific papers with the underlying data. The data discovery function is rather self-explanatory, because it is a means for identifying, reaching, accessing, and retrieving data. No less important and straightforward is citations' function of enabling data sharing, effectuated by data discovery. There can be no doubt that citing data "allows us to define new usage metrics for determining the impact of data." Data citation also can have a positive impact on the reproducibility of science. Earlier attempts to require that citations of data become part of the reference list section in published works show the importance and feasibility of citing research data that can indicate the value of data, because researchers often base their own studies upon data created by others. Being cited is an indicator of influence upon a discipline, and may serve as a measure of reuse by other researchers. Data citations function similarly to traditional citations in academic texts, providing a way to acknowledge intellectual debt to scholarly forebears (Silvello, 2018, pp. 6–10). Two of the best-known providers of data citation indicators are DataCite (https://datacite.org/) and the Data Citation Index (the latter run by Clarivate Analytics). Data citation as an activity within research data management is addressed in Chapter 4.

Data citations of datasets are not the only means for judging the impact of research data and measuring researchers' achievement and prestige. Mapping researchers' influence on the internet has been of concern since at least the late 1990s. However, the concept of altmetrics that covers alternative indicators appeared with the Altmetrics Manifesto (Priem, Taraborelli, Groth, & Neylon, 2010).

Piwowar and Priem (2013, p. 11) discussed 10 benefits of embedding altmetric indicators into someone's CV, because they

- Provide additional information,
- Deemphasize inappropriate metrics,
- Uncover the impact of just-published work,
- Legitimize all types of scholarly products,
- Recognize diverse impact flavors,

- Reward effective efforts to facilitate reuse,
- Encourage a focus on public engagement,
- Facilitate qualitative exploration,
- Empower publication choice, and
- Stimulate the innovation of research evaluation.

Altmetrics are derived from social media, and are distinct from references in formally published, peer-reviewed literature, be it analog or digital. The sources of altmetrics indicators include social media sites, expert reviews and recommendations, social bookmarking sites, gray literature citations, mainstream media, and research and general interest blogs. However, the diversity of data represented under this umbrella term can mean very different things depending on who is using them, what they are doing with them, and how they interpret them (Thelwall & Kousha, 2015). The developments of data publication and the rise of altmetrics are drawing attention to data, and can help researchers understand their influence and share it with others. Altmetrics provides new approaches and measures that draw attention to data and can help researchers understand their influence and share it with others.

Altmetrics indicators for research outputs are designed to provide solutions to assessing the societal impacts of research and obtaining early impact evidence, derived from social media, and different from references that we find in formally published, peer-reviewed literature, be it analog or digital. The sources of altmetrics indicators include social media sites, expert reviews and recommendations, social bookmarking sites, citations in the gray literature and mainstream media, research blogs, and blogs of general interest. However, the diversity of data represented under this umbrella term can mean very different things depending on who is using them, what they are doing with them, and how they interpret them (Thelwall, 2020).

Altmetric indicators are regarded as reflecting engagement with research much earlier than citations and are typically interpreted as a proxy for social impact, attention, and reach.

Usage statistics can indicate overall interest in research data (Lowenberg, Chodacki, Fenner, Kemp, & Jones, 2019). Usage statistics are often reported through a standard called COUNTER, which is capable of ensuring consistency of tracking and reporting (COUNTER, 2014).

Reuse metrics can showcase the scholarly influence of research data and the diversity of contexts in which data is reused, but there is uncertainty as to whether the instances of data use they point to are substantive or not. In addition to this, they do not directly measure reuse, and do not allow judging the value of reuse.

Though altmetrics are promising, there are limitations to their use in research data evaluation. The biggest challenge is that altmetrics indicators are not yet widely used in research assessment, and thus the literacy surrounding their interpretation and responsible use is low. Another problematic issue is that online attention for research can be fabricated easily. In addition, altmetrics aggregators cannot yet offer useful disciplinary benchmarking for research data performance, due to the limited number of repositories that are indexed by them (Konkiel, 2020).

Although full of promise, altmetrics are imperfect for a number of reasons. They do not meet the requirements for data quality, especially because they may be incorrect as data sources. Few altmetrics data sources normalize their data, so cross-discipline comparisons are difficult, and the available tools are not transparent regarding data coverage. Moreover, altmetrics are subject to general and disciplinary biases. They may be incorrect, and often do not measure impact but rather the extent of attention. Just like citation counts, they cannot be properly interpreted if they are used in isolation.

To overcome these difficulties, it is recommended to use altmetrics always in context. Altmetrics indicators themselves cannot serve as evidence. Neither are they equal to evaluative mechanisms or to sets of required benchmarks, and therefore cannot and should not be imposed on researchers.

It is advisable thus to align research data impact indicators with the principles articulated in the Leiden Manifesto for research metrics, which states that evaluation should be done by subject experts, equipped to assess data quality and its impact on its own merits. On the other hand, measurements of performance should take the research missions of the institution, group, or researcher into account. There is a need to do the same for variations of publication and citation practices, as well as in regard to the strengths and limitations of the data. The Manifesto advocates that metrics providers and their users should show transparency in their data collection and aggregation processes (Hicks, Wouters, Waltman, De Rijcke, & Rafols, 2015).

Although data quality has been addressed in another chapter of this book in more detail, some of its dimensions and attributes must be explained here as well. Quality metrics may vary in different disciplines and data structures. This should be considered when setting standards for assessing research data. Moreover, due to the lack of standardization, displaying research impact by data citation practices has limited success. We see a similar situation in the case of usage statistics and reuse metrics. The availability of altmetrics for research data is very limited. Although supported by researchers, using numeric indicators is not widely practiced yet. Reuse metrics can highlight the scholarly influence of research data and the diversity of contexts in which data is reused. Unfortunately, there is uncertainty as to whether the instances of data use are substantive or not. In addition to this, they do not directly measure reuse, nor do they allow judging the value of reuse.

While there are numerous varieties of metrics, such as quality indicators, citation-based indicators, altmetrics, usage statistics, and reuse indicators, altmetrics cover only a portion of the available ones. Moreover, measuring quality must be ongoing and iterative (Konkiel, 2020).

The case of Publons (https://publons.com) represents a third approach, because it is similar to other academic social networking services (ASNSs) that are discussed in some detail in a following section. It also provides various alternative scholarly metrics and an overall score. However, Publons offers rewards based on the number of papers that the given researcher has peer-reviewed (Smith, 2016). Taking into account that peer-review activity is an independent facet of scholarly activity, this implies that correlations between these activities are weak by their differing nature (Ortega, 2017).

Put into a broader context, research evaluation in harmony with the goals of open science should be based on 23 criteria, grouped into 6 categories. Research output includes research activities, (open source and other) publications, produced datasets, and funding. The research process should be appraised by stakeholder engagement and/or by involving citizen science, collaboration and interdisciplinarity, the integrity of research, and risk management. Service and leadership involve not only leadership itself but also academic standing, networking, and participating in peer review. Research impact consists of communicating and disseminating, patents and licenses, societal impact, and knowledge exchange. Teaching and supervising fall into the categories of teaching, mentoring, and supervision. Professional experience includes continuing professional development, project management, and developing personal qualities (Rentier, 2019). Obviously, some of the requirements cannot be fulfilled due to disciplinary differences. Let us only mention that researchers in the social sciences or in the humanities rarely, or hardly ever, produce patents or licenses.

Other issues of the information ecosystem

A number of relatively divergent issues related to information and data are not directly tied to research data, but must be dealt with due to their role in the ecosystem.

Research information management

Academic libraries can play a leadership and advocacy role in the organizational discussion around research information management (RIM) (Materska, 2019). RIM is a relatively new type among the types of service that may require the involvement of the library, but it is different from research data management (Day, 2017). RIM refers to the "management, evaluation, and disclosure of research outcomes and expertise, which connects in various ways with internal evaluation and management goals, funding policy and compliance needs, as well as with broader reputation management on the web" (Dempsey, 2017, p. 344). These systems are also called research information management systems (RIMS), or current research information systems (CRIS). They facilitate "uniting information from the different institutional systems under a common interface as they allow an organization to achieve the full array of life-cycle needs, from the identification of funding, through reporting and benchmarking of research outputs, to high level strategic planning" (Day, 2017, p. 24).

RIMS enable managing and providing access to researchers' authored content and identity information. Research identity may include "not only the publications, datasets, and research technologies produced by researchers, but information about the researcher's capabilities, skills, and expertise, as well." Part of these, i.e., publication information, can be collected automatically, while information on other facets of research identity may not be readily available (Stvilia, Wu, & Lee, 2018, p. 1035).

Reputation management

Disclosing university assets is an important activity in the context of institutional reputation. This can be directed toward the "institution's unique intellectual products that include archives and special collections, or newly generated research and learning materials (e-prints, research data, courseware, digital scholarly resources, etc.), or such things as expertise or researcher profiles." It is often the goal to share these materials with potential users outside the institution. Beyond all this, there is a growing interest in discovering researcher profiles as well, because it makes their achievements and publications more readily discoverable (Dempsey, 2017, p. 341).

With the advent and spread of social networking, a number of researchers are inevitably establishing their presence in academic social networking services (ASNSs), which are often perceived as effective tools for the discovery and dissemination of research outputs (Greifeneder et al., 2018). As uploading full texts generates usage and discussion, it is the preferred form of participating in these ASNSs. Participants receive messages about reads and citations of their publications and there are followers, who play a similar role to those in other social networks (Nicholas, Clark, & Herman, 2016). The following short, nonexhaustive review of these ASNSs demonstrates some of their features.

ResearchGate (RG) (https://www.researchgate.net/) is one of the best-known among these services. It enables researchers to disseminate and share scholarly publications free of charge all over the world

(Yu, Wu, Alhalabi, Kao, & Wu, 2016). RG also offers a number of possibilities to build up, display, and measure the reputations of the researchers. Its notable service is the ability to include descriptions of the researcher's ongoing projects (Hubbe, 2017).

In principle, uploading papers to ASNSs should abide by the rules of intellectual property rights. Nonetheless, this is not always the case.

Jamali (2017, pp. 241–242) examined a random sample of journal articles available as full text on ResearchGate. The results show that, out of these 500 papers, 21.6% were open access. Of the remaining 392 articles, 15.6% were preprint, 6.1% postprint, and 78.3% were published PDF (often called publishers' PDF files, or final published versions). The key finding was that "51.3% of non-OA articles infringed copyright regulations and were noncompliant with publishers' policies." ResearchGate itself "does not check uploaded files to see if they are uploaded legally, and in compliance with copyright; and not all authors comply with their agreements with journal publishers." This indicates that authors might lack understanding of copyright, or do not care enough about this issue. RG also contributes to reputation management, discussed in the subsection of the same name, by providing its own score.

Academia.edu (https://www.academia.edu/) fulfills a similar function. While many researchers also can receive publications free of charge through it, some of its services, e.g., alerts about citations, are behind a paywall. At the end of 2020, they began experimenting with broadening their portfolio by publishing a new online journal called *Academia Letters* that contains rapid, peer-reviewed, short (800–1600 words) papers. Elsevier's Mendeley (https://www.mendeley.com) is not only an SNS but a free reference manager as well. Kudos (https://growkudos.com) is also distinct because it is supported by several commercial journal publishers, and it is intended to serve as an intermediary among authors of scholarly articles, academic institutions, publishers, and the general public through simplifying and interpreting these papers (Williams, 2017).

Many of these ASNSs provide altmetrics indicators, mentioned in the following section on measuring and evaluating research data.

Beyond the features described, reputation management has begun to focus increasingly on research data. The Hungarian Scientific Bibliography (MTMT, https://www.mtmt.hu/) database is one of the simple examples that show this development. While this database registers the bibliographic data of Hungarian researchers' publications and the citations to these publications, it also looks for research data by offering a dedicated field where datasets can be described. On an international level, Mendeley offers a free tool for research data management, called Mendeley Data (https://data.mendeley.com/).

Posting publicly available versions of scientific articles to dedicated websites prior to undergoing the traditional peer-review process, i.e., preprints, is not a new idea (Sarabipour et al., 2019). We know that physicists have been using a preprint system to share and discuss their work since the early 1990s. This system has become arXiv.org (https://arxiv.org/), which is currently the largest preprint server in the world. It holds preprints from a number of many disciplines, including physics, mathematics, computer science, finance, statistics, economics, and engineering. A preprint server dedicated to the biological sciences, bioRxiv (https://biorxiv.org/), came online in 2013, while MedRxiv (https://medrxiv.org/), serving clinical research, was launched in 2019. There are preprint servers for other disciplines as well. OSFPreprints (https://osf.io/preprints/) was funded by the Open Science Foundation (https://osf.io/). It provides infrastructure to smaller organizations in a variety of disciplines including architecture, art, business, education, engineering, law, life sciences, medicine, mathematics, and social sciences.

Posting preprints exposes works to a much larger group than the traditional peer-review process. Preprints expose articles that would otherwise not be published. Less frequently published issues, such as negative results, replications, and contradictions, may be showcased as preprints. Moreover, they provide free public versions of articles that are "very close to the final published version. For researchers with limited access to a library, preprints can fill an important access gap," among other things, by helping to overcome paywalls (Hoy, 2020, p. 86).

Demonstrating value by coauthoring publications

Academic libraries are increasingly called upon to demonstrate their value beyond service functions and LIS to their home institutions. Borrego, Ardanuy, and Urbano (2018) provide evidence of collaboration between librarians and research groups in the form of coauthored publications. Such collaborations between librarians and other researchers seem to exemplify "the embedded librarian model in working directly with the faculty they serve as collaborators on research projects or as an integral part of a research team" (Carlson & Kneale, 2011).

An international survey by Borrego and Pinfield (2020) garnered 60 responses from librarians, who can be regarded as coauthors of scholarly articles in journals that are outside library and information science. In this sample, 43% of the participants had a PhD, which is possibly higher than the average share of librarians in general. Nonetheless, librarians' contribution was substantial. Not surprisingly, 22% of the respondents were involved in designing and conducting the literature search, managing the references, and writing the methodological section of the paper. Many of them were also reviewing and editing the final publication. A higher number of respondents (67%) referred to fields of expertise that are typically tied to the library profession, i.e., conducting literature reviews, selecting databases, defining search strategies, using reference management software, and using specific software for systematic reviews.

There were perceived benefits that influence not only research teams. Collaboration allowed librarians to better understand researchers' needs and behavior, as well as improve research support services. Conducting research benefited librarians also "by improving their problem-solving and decision-making skills and by making them critical consumers of academic literature not limited to their own profession. Carrying out research in partnership with faculty also seems to have a potential to offer librarians opportunities to gain valuable experience in how research operates" (p. 656).

Information and data overload

Information overload (IO) is a set of objective and subjective difficulties caused by the amount and complexity of information available, as well as by our inability to handle such situations (Koltay, 2021). IO is by no means a myth, especially if we look at the history of recorded information that showed most of the time a perception of overload.

Overload is caused not only by too much information. Its diversity, complexity, and novelty also can become an issue (Bawden & Robinson, 2020). Besides information overload, due to the lack of substantial barriers to data flow, we can also feel the symptoms of data overload (Virkus, Mandre, & Pals, 2018). There is also communication overload, caused by overcommunication and miscommunication, among other factors (Lehman & Miller, 2020). Despite differences between them, it is reasonable to label all overload phenomena as *information overload*.

Our information environment is characterized not only by larger amounts of information and a greater variety of its formats and types available. However, TMI (Too Much Information) is not the only cause of information overload (Bawden & Robinson, 2009). Due to a lack of professional support from outside of their teams (e.g., by academic librarians), researchers may not utilize personalized digital research tools properly, and thus they may become unable to ease the burden of information overload (Nolin, 2013). In the case of data, we also face the problem of "finding the significance of data when it is not known a priori what data from a large data field will be informative" (Woods, Patterson, & Roth, 2002, p. 25).

Information overload appears on two levels. Its macro level is related to the limits of physical storage and (mainly) human processing capacities in managing information properly. The micro level is of a qualitative nature, as it is essentially a failure to filter out unnecessary, not wanted, surplus information (Davis, 2011).

Mitigating the symptoms of IO can be based on varied strategies. Some of them are of a technical (technological) nature, while others tackle social issues. Notwithstanding, in many cases they represent two sides of the same coin (Benselin & Ragsdell, 2016).

From among technological approaches, information architecture (IA) may play a distinguished role, as its main responsibility is designing, implementing, and evaluating acceptable and usable information spaces. However, the power of IA in this regard is generally limited, because information architects can discover information overload only retrospectively and through indirect means (Davis, 2012).

Keeping and discarding information based on using critical assessment may be a general solution to information overload, and—contrary to technological solutions—social approaches offer a much broader repertoire of answers. This is especially true for the micro level of information overload, propagated by people and caused by the lack of adequate filters to information or failure to apply them properly. This type of overload can be combated by offering appropriate education that focuses on critical thinking, which is part of information literacy as it offers frameworks for reducing overload (Blummer & Kenton, 2014). Data literacy education has a similar effect, especially if we are selective about data sources and focus more on qualitative data (Saxena & Lamest, 2018).

Personal information management (PIM) strategies related to these literacies mean, broadly speaking, taking control of someone's information environment. These strategies gain special weight for treating IO, because these strategies can help researchers to recognize their information needs more explicitly. Being aware of them substantiates decision-making for storing and/or discarding information held personally (Bergman, 2013). IO is often applied to research information collections, which come into existence by a conscious effort to control the quantity and quality of incoming information and organize it accordingly (Bussert, Chiang, & Tancheva, 2011).

Abstracts are rarely mentioned as related to information overload, and abstracting means representing (selectively summarizing) the most important content of an already existing (professional or scholarly) text (mainly a journal paper) into self-contained texts that incorporate this information in a concise, condensed, and abbreviated, thus shorter, form than the primary (original) text. This can be performed by the same authors who originally produced these texts, or by professional abstractors, who are usually commissioned by some organization for this task, and who thus differ from the author(s) of the original text (Koltay, 2010).

Although originally restricted to textual information, they are an important part of the scholarly communication. Moreover, summarization has also reached the word of data. Summarizing data is a means of making sense of it, improving therefore data discovery (Koesten, Simperl, Blount, Kacprzak, & Tennison, 2020). Data summarization is applied to both structured and unstructured data by making use of machine learning, statistical analyses, semantics, and natural language processing, among others (Ahmed, 2019).

As described by Ojala (2010), in some cases abstracts have been judged anachronistic. In contrast to such opinions, this genre has remained actual as the use of abstracts in research and applying them to the production of information is vital. Abstracting must be based on concentrating on the importance of certain, selected parts of the information and/or data, by making use of critical thinking that is an essential constituent of information and data literacy. The content of scholarly papers is reflected in the abstracts, depending on what is considered important by the author and by the abstractor in order to facilitate judging its relevance to a particular information need. Scientific abstracts therefore reproduce only part of the information and the complexity of argumentation in a scientific article (Atanassova, Bertin, & Larivière, 2016).

According to Izquierdo Alonso and Moreno Fernandez (2010), professional abstractors have to consider their mediating role, because it requires rhetorical, functional, and semantic choices. Abstracting unfolds in specialized contexts by attending to a series of contextual probabilities, which define the level of adopting the original text's content, seeking equivalence, and applying heterovalent adaptation, based on the perceived expectations of the discourse community. These expectations require applying certain patterns of sociocognitive informative and communicative behavior.

In order to avoid misunderstandings, it must be added that the word *abstract* is also often used as a partial homonym, because it may mean a short document produced by someone aspiring to be accepted as a speaker at a conference. Such conference abstracts are submitted at the time of applying to a conference, and thus they do not reflect an existing text but are promissory by foretelling what the applicant intends to speak about.

Although there has been a longstanding interest in automatic abstracting (text summarization) beginning from the early days of computational linguistics (Orasan, 2019), and the activities are continually researched today, abstracting has nonetheless remained mainly an intellectual (manual) labor.

Facing posttruth phenomena and the COVID-19 pandemic

Besides the presence of information overload, the proliferation of misinformation and disinformation hampers the consumption and use of information and data as sources of knowledge and understanding. The prevalence of posttruth phenomena such as fake news, misinformation, and disinformation aggravates the overload situation. There is accordingly a need for separating useful information and data from the useless ones by verifying their provenance and truthfulness.

In the case of scholarly publications, critical thinking is aided by peer review. Nonetheless, information literacy and data literacy may foster critical reading that is based on critical thinking by distinguishing between content and its presentation (i.e., purpose, central claims, intended audience, evidence, and sources). Although researchers are supposed to adhere to evidence and strict rules of inference, it may happen that they do not always invest their full mental capacities in the tasks of information acquisition and evaluation (Materska, 2014).

Taking time for deconstructing messages may offer a useful approach for differentiating true information from information that is false or misleading. Deconstruction is a semiotic analysis that focuses on language, genres, authorship, format, and audience. It is based on the premise that individuals should not simply perceive textual or media messages as neutral, but must recognize the constructed nature of these messages. Reconstructing is also needed, for example in the form of comparing the information about the same news from different types of media and/or the (political, ideological, etc.) commitment of the given media outlet that was used to construct the messages (Lin, Li, Deng, & Lee, 2013).

The COVID-19 pandemic led to a deluge of related publications. The growth in the number of preprints (mentioned in one of the preceding sections) is especially spectacular in this field. In a cross-sectional study, the number and reporting characteristics of COVID-19-related academic articles and preprints was compared. Searches in the PubMed database as well as in MedRxiv and BioRxiv archives showed that the number of academic and preprint publications increased significantly over time, but the latter were slightly more prevalent than academic articles. This shows that "preprints have gained traction as a publishing response to the demand for prompt access to empirical albeit not peer-reviewed, findings during the present pandemic" (Gianola, Jesus, Bargeri, & Castellini, 2020, p. 1). The findings of Majumder and Mandl (2020) also suggest that, thanks to the speed of their quick release, preprints might be the drivers of the discourse around COVID-19.

References

Ahmed, M. (2019). Data summarization: A survey. *Knowledge and Information Systems*, *58*(2), 249–273. https://doi.org/10.1007/s10115-018-1183-0.

Atanassova, I., Bertin, M., & Larivière, V. (2016). On the composition of scientific abstracts. *Journal of Documentation*, *72*(4), 636–647. https://doi.org/10.1108/JDOC-09-2015-0111.

Ayris, P. (2021). "The future depends on what you do today": The library as a leader in open science. In *Cases on research support services in academic libraries* (pp. 25–51). Hershey, PA: IGI Global. https://doi.org/10.4018/978-1-7998-4546-1.ch002.

Bates, J. (2018). The politics of data friction. *Journal of Documentation*, *74*(2), 412–429. https://doi.org/10.1108/JD-05-2017-0080.

Bawden, D., & Robinson, L. (2009). The dark side of information: Overload: Anxiety and other pathologies. *Journal of Information Science, 35*(2), 180–191. https://doi.org/10.1177/0165551508095781.

Bawden, D., & Robinson, L. (2020). Information overload: An introduction. In *Oxford encyclopedia of political decision making*. Oxford: Oxford University Press. https://doi.org/10.1093/acrefore/9780190228637.013.1360.

Benedetti, A., Boehme, G., Caswell, T. R., Denlinger, K., Li, Y., McAllister, A. D., ... Wesolek, A. J. (2020). 2020 top trends in academic libraries. *College & Research Libraries News, 81*(6), 270–278. https://doi.org/10.5860/crln.81.6.270.

Benselin, J. C., & Ragsdell, G. (2016). Information overload: The differences that age makes. *Journal of Librarianship and Information Science, 48*(3), 284–297. https://doi.org/10.1177/961000614566341.

Bergman, O. (2013). Variables for personal information management research. *ASLIB Proceedings, 65*(5), 464–483. https://doi.org/10.1108/AP-04-2013-0032.

Blanchette, J.-F. (2012). *Burdens of proof: Cryptographic culture and evidence law in the age of electronic documents*. Cambridge, MA: The MIT Press.

Blummer, B., & Kenton, J. M. (2014). Reducing patron information overload in academic libraries. *College & Undergraduate Libraries, 21*(2), 115–135. https://doi.org/10.1080/10691316.2014.906786.

Borgman, C. L. (2012). The conundrum of sharing research data. *Journal of the American Society for Information Science and Technology, 63*(6), 1059–1078. https://doi.org/10.1002/asi.22634.

Borgman, C. L. (2015). *Big data, little data, no data: Scholarship in the networked world*. Cambridge, MA: MIT Press.

Borrego, Á., Ardanuy, J., & Urbano, C. (2018). Librarians as research partners: Their contribution to the scholarly endeavour beyond library and information science. *The Journal of Academic Librarianship, 44*(5), 663–670. https://doi.org/10.1016/j.acalib.2018.07.012.

Borrego, Á., & Pinfield, S. (2020). Librarians publishing in partnership with other researchers: Roles, motivations, benefits, and challenges. *Portal: Libraries and the Academy, 20*(4), 655–675. https://doi.org/10.1353/pla.2020.0031.

Buckland, M. (2011). Data management as bibliography. *Bulletin of the American Society for Information Science and Technology, 37*(6), 34–37. https://doi.org/10.1002/bult.2011.1720370611.

Bugaje, M., & Chowdhury, G. (2018). Identifying design requirements of a user-centered research data management system. In M. Dobreva, A. Hinze, & M. Žumer (Eds.), *Maturity and innovation in digital libraries* (pp. 335–347). Cham: Springer. https://doi.org/10.1007/978-3-030-04257-8_35.

Burton, M., & Lyon, L. (2017). Data science in libraries. *Bulletin of the Association for Information Science and Technology, 43*(4), 33–35. https://doi.org/10.1002/bul2.2017.1720430409.

Bussert, K., Chiang, K., Tancheva, K., Foster, N. F., Clark, K., Tancheva, K., & Kilzer, R. (2011). Personal management of scholarly information. In *Scholarly practice, participatory design and the eXtensible catalog* (pp. 123–151). Chicago, IL: Association of College and Research Libraries (chapter 6).

Candela, L., Castelli, D., Manghi, P., & Tani, A. (2015). Data journals: A survey. *Journal of the Association for Information Science and Technology, 66*(9), 1747–1762. https://doi.org/10.1002/asi.23358.

Carlson, J., & Kneale, R. (2011). Embedded librarianship in the research context: Navigating new waters. *College & Research Libraries News, 72*(3), 167–170. https://doi.org/10.5860/crln.72.3.8530.

Chawinga, W. D., & Zinn, S. (2020). Research data management in universities: A comparative study from the perspectives of librarians and management. *The International Information and Library Review*. https://doi.org/10.1080/10572317.2020.1793448.

Copeland, A., Yoon, A., & Zhang, S. (2020). Data reuse practices and expectations for data resources and services among public library users. *Public Library Quarterly*. https://doi.org/10.1080/01616846.2020.1773749.

COUNTER. (2014). *COUNTER about us*. http://www.projectcounter.org/about.html.

Davis, N. (2011). Information overload, reloaded. *Bulletin of the American Society for Information Science and Technology*, *37*(5), 45–49. https://doi.org/10.1002/bult.2011.1720370513.

Davis, N. (2012). *IA strategy: Addressing the signatures of information overload*. UXmatters. February 2012. Retrieved from http://www.uxmatters.com/mt/archives/2012/02/ia-strategy-addressing-the-signatures-of-information-overload.php.

Day, A. P. (2017). Research information management: How the library can contribute to the campus conversation. *New Review of Academic Librarianship*, *24*(1), 23–34. https://doi.org/10.1080/13614533.2017.1333014.

Dempsey, L. (2017). Library collections in the life of the user: Two directions. *LIBER Quarterly*, *26*(4), 338–359. https://doi.org/10.18352/lq.10170.

Drobíková, B., Jarolímková, A., & Souček, M. (2017). Data literacy among Charles University PhD students: Are they prepared for their research careers? In *European conference on information literacy* (pp. 169–177). Cham: Springer. https://doi.org/10.1007/978-3-319-74334-9_18.

EC (n.d.). *European open science cloud (EOSC)*. Bruxelles: European Commission. Retrieved from https://ec.europa.eu/info/research-and-innovation/strategy/goals-research-and-innovation-policy/open-science/european-open-science-cloud-eosc_en.

Edwards, P. N., Mayernik, M. S., Batcheller, A. L., Bowker, G. C., & Borgman, C. L. (2011). Science friction: Data, metadata, and collaboration. *Social Studies of Science*, *41*(5), 667–690. https://doi.org/10.1177/0306312711413314.

Faniel, I. M., & Connaway, L. S. (2018). Librarians' perspectives on the factors influencing research data management programs. *College & Research Libraries*, *79*(1), 100–119. https://doi.org/10.5860/crl.79.1.100.

Faniel, I. M., & Jacobsen, T. E. (2010). Reusing scientific data: How earthquake engineering researchers assess the reusability of colleagues' data. *Computer Supported Cooperative Work*, *19*(3), 355–375. https://doi.org/10.1007/s10606-010-9117-8.

Faniel, I. M., Kriesberg, A., & Yakel, E. (2016). Social scientists' satisfaction with data reuse. *Journal of the Association for Information Science and Technology*, *67*(6), 1404–1416. https://doi.org/10.1002/asi.23480.

Federer, L. (2016). Research data management in the age of big data: Roles and opportunities for librarians. *Information Services & Use*, *36*(1/2), 35–43.

Floridi, L., & Taddeo, M. (2016). What is data ethics? *Philosophical Transactions of the Royal Society A*, *374*, 20160360. https://doi.org/10.1098/rsta.2016.0360.

Gianola, S., Jesus, T. S., Bargeri, S., & Castellini, G. (2020). Characteristics of academic publications, preprints, and registered clinical trials on the COVID-19 pandemic. *PLoS One*, *15*(10), e0240123. https://doi.org/10.1371/journal.pone.024012.

Gold, E. R., Ali-Khan, S. E., Allen, L., Ballell, L., Barral-Netto, M., Carr, D., … Cook-Deegan, R. (2019). An open toolkit for tracking open science partnership implementation and impact. *Gates Open Research*, *3*, 1442. https://doi.org/10.12688/gatesopenres.12958.2.

Greifeneder, E., Pontis, S., Blandford, A., Attalla, H., Neal, D., & Schlebbe, K. (2018). Researchers' attitudes towards the use of social networking sites. *Journal of Documentation*, *74*(1), 119–136. https://doi.org/10.1108/JD-04-2017-0051.

Haendel, M. A., Vasilevsky, N. A., & Wirz, J. A. (2012). Dealing with data: A case study on information and data management literacy. *PLoS Biology*, *10*(5), e1001339. https://doi.org/10.1371/journal.pbio.1001339.

Hicks, D., Wouters, P., Waltman, L., De Rijcke, S., & Rafols, I. (2015). Bibliometrics: The Leiden manifesto for research metrics. *Nature News*, *520*(7548), 429–430.

Hoy, M. B. (2020). Rise of the Rxivs: How preprint servers are changing the publishing process. *Medical Reference Services Quarterly, 39*(1), 84–89. https://doi.org/10.1080/02763869.2020.1704597.

Hrynaszkiewicz, I. (2019). Publishers' responsibilities in promoting data quality and reproducibility. In A. Bespalov, M. Michel, & T. Steckle (Eds.), *Handbook of experimental pharmacology*. Cham: Springer.

Huang, H. (2018). Big data to knowledge—Harnessing semiotic relationships of data quality and skills in genome curation work. *Journal of Information Science, 44*(6), 785–801. https://doi.org/10.1177/0165551517748291.

Hubbe, M. A. (2017). Why I don't do academic social media… or do I? *BioResources, 12*(2), 2252–2253.

Hyndman, A., & Goodey, G. (2020). It all starts with a good plan. In *The state of open data* (pp. 17–21). Cham: Springer Nature. https://doi.org/10.6084/m9.figshare.13227875.

Izquierdo Alonso, M., & Moreno Fernandez, L. M. (2010). Perspectives of studies on document abstracting: Towards an integrated view of models and theoretical approaches. *Journal of Documentation, 66*(4), 563–584. https://doi.org/10.1108/00220411011052957.

Jamali, H. R. (2017). Copyright compliance and infringement in ResearchGate full-text journal articles. *Scientometrics, 112*(1), 241–254. https://doi.org/10.1007/s11192-017-2291-4.

Jávorszky, F., Murányi, P., & Koltay, T. (2019). A magyar kutatók adatműveltségi készségeiről. *Könyvtári Figyelő, 29*(3), 363–374 (in Hungarian).

Jiao, C., & Darch, P. T. (2020). The role of the data paper in scholarly communication. *Proceedings of the Association for Information Science and Technology, 57*(1), e316. https://doi.org/10.1002/pra2.316.

Joo, S., & Peters, C. (2020). User needs assessment for research data services in a research university. *Journal of Librarianship and Information Science, 52*(3), 633–646. https://doi.org/10.1177/0961000619856073.

Kim, J. (2020). An analysis of data paper templates and guidelines: Types of contextual information described by data journals. *Science Editing, 7*(1), 16–23. https://doi.org/10.6087/kcse.185.

Kim, Y., & Yoon, A. (2017). Scientists' data reuse behaviors: A multilevel analysis. *Journal of the Association for Information Science and Technology, 68*(12), 2709–2719. https://doi.org/10.1002/asi.23892.

Kirkwood, R. J. (2016). Collection development or data-driven content curation? *Library Management, 37*(4/5), 275–284. https://doi.org/10.1108/LM-05-2016-0044.

Klain Gabbay, L., & Shoham, S. (2019). The role of academic libraries in research and teaching. *Journal of Librarianship and Information Science, 51*(3), 721–736. https://doi.org/10.1177/0961000617742462.

Koers, H., Bangert, D., Hermans, E., van Horik, R., de Jong, M., & Mokrane, M. (2020). Recommendations for services in a FAIR data ecosystem. *Patterns*, 100058. https://doi.org/10.1016/j.patter.2020.100058.

Koesten, L., Simperl, E., Blount, T., Kacprzak, E., & Tennison, J. (2020). Everything you always wanted to know about a dataset: Studies in data summarisation. *International Journal of Human-Computer Studies, 135*, 102367. https://doi.org/10.1016/j.ijhcs.2019.10.004.

Koltay, T. (2010). *Abstracts and abstracting: A genre and set of skills for the twenty-first century*. Oxford: Chandos.

Koltay, T. (2021). Information overload. In D. B. A. Khosrow-Pour (Ed.), *Encyclopedia of organizational knowledge, administration, and technology* (pp. 1805–1816). Hershey, PA: IGI Global.

Konkiel, S. (2020). Assessing the impact and quality of research data using altmetrics and other indicators. *Scholarly Assessment Reports, 2*(1), 13. https://doi.org/10.29024/sar.13.

Lehman, A., & Miller, S. J. (2020). A theoretical conversation about responses to information overload. *Information, 11*(8), 379. https://doi.org/10.3390/info11080379.

LERU. (2018). *Open science and its role in universities. A roadmap for cultural change*. Retrieved from https://www.leru.org/files/LERU-AP24-Open-Science-full-paper.pdf.

Lin, T.-B., Li, J.-Y., Deng, F., & Lee, L. (2013). Understanding new media literacy: An explorative theoretical framework. *Educational Technology & Society*, *16*(4), 160–170.

Lowenberg, D., Chodacki, J., Fenner, M., Kemp, J., & Jones, M. B. (2019). *Open data metrics: Lighting the fire*. https://doi.org/10.5281/zenodo.3525349.

Majumder, M. A., & McGuire, A. L. (2020). Data sharing in the context of health-related citizen science. *The Journal of Law, Medicine, & Ethics*, *48*(1_Suppl), 167–177. https://doi.org/10.1177/1073110520917044.

Majumder, M. S., & Mandl, K. D. (2020). Early in the epidemic: Impact of preprints on global discourse about COVID-19 transmissibility. *The Lancet Global Health*, *8*(5), e627–e630. https://doi.org/10.1016/S2214-109X(20)30113-3.

Materska, K. (2014). Information heuristics of information literate people. In S. Kurbanoğlu, S. Špiranec, E. Grassian, D. Mizrachi, & L. Catts (Eds.), *Information literacy. Lifelong learning and digital citizenship in the 21st century* (pp. 59–69). Cham: Springer.

Materska, K. (2019). Research information management in the context of open science. *Zagadnienia Informacji Naukowej–Studia Informacyjne*, *57*(1A), 37–45.

Mons, B. (2018). *Data stewardship for open science. Implementing FAIR principles*. New York: Chapman and Hall/CRC.

Nafus, D. (2014). Stuck data, dead data, and disloyal data: The stops and starts in making numbers into social practices. *Distinktion: Scandinavian Journal of Social Theory*, *15*(2), 208–222. https://doi.org/10.1080/1600910X.2014.920266.

Neylon, C., Belso, R., Bijsterbosch, M., Cordewener, B., Foncel, J., Frieske, S., … Sesink, L. (2019). *Open scholarship and the need for collective action*. Knowledge Exchange.

Nicholas, D., Clark, D., & Herman, E. (2016). ResearchGate: Reputation uncovered. *Learned Publishing*, *29*(3), 173–182. https://doi.org/10.1002/leap.1035.

Nolin, J. M. (2013). The special librarian and personalized meta-services: Strategies for reconnecting librarians and researchers. *Library Review*, *62*(8/9), 508–524. https://doi.org/10.1108/LR-02-2013-0015.

Ogburn, J. L. (2010). The imperative for data curation. *portal: Libraries and the Academy*, *10*(2), 241–246. https://doi.org/10.1353/pla.0.0100.

Ojala, M. (2010). An abstract concept. *Online*, *36*(2), 5. Retrieved from http://www.infotoday.com/online/mar12/HomePage-An-Abstract-Concept.shtml.

Orasan, C. (2019). Automatic summarisation: 25 years on. *Natural Language Engineering*, *25*(6), 735–751.

Ortega, J. L. (2017). Are peer-review activities related to reviewer bibliometric performance? A scientometric analysis of Publons. *Scientometrics*, *112*, 947–962. https://doi.org/10.1007/s11192-017-2399-6.

Oßwald, A. (2021). Barrieren, hemmschwellen und gatekeeper. In M. Putnings, H. Neuroth, & J. Neumann (Eds.), *Praxishandbuch Forschungsdatenmanagement* (pp. 277–279). Berlin: De Gruyter, SAUR (in German).

Pálsdóttir, Á. (2017). Data literacy, collaboration and sharing of research data among academics at the University of Iceland. In *European conference on information literacy* (pp. 178–185). Cham: Springer. https://doi.org/10.1007/978-3-319-74334-9_19.

Pálsdóttir, A. (2021). Data literacy and management of research data—A prerequisite for the sharing of research data. *Aslib Journal of Information Management*. https://doi.org/10.1108/AJIM-04-2020-0110.

Pangrazio, L., & Sefton-Green, J. (2020). The social utility of 'data literacy'. *Learning, Media and Technology*, *45*(2), 208–220. https://doi.org/10.1080/17439884.2020.1707223.

Pasquetto, I. V., Randles, B. M., & Borgman, C. L. (2017). On the reuse of scientific data. *Data Science Journal*, *16*(8). https://doi.org/10.5334/dsj-2017-008.

Perrier, L., Blondal, E., & MacDonald, H. (2020). The views, perspectives, and experiences of academic researchers with data sharing and reuse: A meta-synthesis. *PLoS One*, *15*(2), e0229182. https://doi.org/10.1371/journal.pone.0229182.

Piwowar, H., & Priem, J. (2013). The power of altmetrics on a CV. *Bulletin of the American Society for Information Science and Technology*, *39*(4), 10–13. https://doi.org/10.1002/bult.2013.1720390405.

Piwowar, H., Priem, J., Larivière, V., Alperin, J. P., Matthias, L., Norlander, B., ... Haustein, S. (2018). The state of OA: A large-scale analysis of the prevalence and impact of open access articles. *PeerJ*, *6*, e4375. https://doi.org/10.7717/peerj.4375.

Poole, A. H. (2016). The conceptual landscape of digital curation. *Journal of Documentation*, *72*(5), 961–986. https://doi.org/10.1108/JD-10-2015-0123.

Priem, J., Taraborelli, D., Groth, P., & Neylon, C. (2010). *Altmetrics: A manifesto*. http://altmetrics.org/manifesto/.

RDA (n.d.). *About RDA*. Research Data Alliance. Retrieved from https://www.rd-alliance.org/about-rda.

Rentier, B. (2019). Open Science will never prevail without a thorough revisiting of the way evaluations of researchers are conducted [Lecture notes]. In *OAI 11, the Cern-Unige workshop on innovations in scholarly communication, Geneva, 19-21 June 2019*. Retrieved from https://indico.cern.ch/event/786048/contributions/3304623/.

Robinson, L., & Bawden, D. (2017). 'The story of data': A socio-technical approach to education for the data librarian role in the CityLIS library school at City, University of London. *Library Management*, *38*(6/7), 312–322. https://doi.org/10.1108/LM-01-2017-0009.

Sarabipour, S., Debat, H. J., Emmott, E., Burgess, S. J., Schwessinger, B., & Hensel, Z. (2019). On the value of preprints: An early career researcher perspective. *PLoS Biology*, *17*(2), e3000151. https://doi.org/10.1371/journal.pbio.3000151.

Savage, J. L., & Cadwallader, L. (2019). Establishing, developing, and sustaining a community of data champions. *Data Science Journal*, *18*(23), 1–8. https://doi.org/10.5334/dsj-2019-023.

Saxena, D., & Lamest, M. (2018). Information overload and coping strategies in the big data context: Evidence from the hospitality sector. *Journal of Information Science*, *44*(3), 287–297. https://doi.org/10.1177/0165551517693712.

Sayre, F., & Riegelman, A. (2018). The reproducibility crisis and academic libraries. *College & Research Libraries*, *79*(1), 2–9. https://doi.org/10.5860/crl.79.1.2.

Schöpfel, J., Farace, D., Prost, H., Zane, A., & Hjørland, B. (2020). Data documents. In *ISKO encyclopedia of knowledge organization*. Retrieved from https://www.isko.org/cyclo/data_documents.htm.

Sewell, C., & Kingsley, D. (2017). Developing the 21st century academic librarian: The Research Support Ambassador Programme. *New Review of Academic Librarianship*, *23*(2–3), 148–158. https://doi.org/10.1080/13614533.2017.1323766.

Silvello, G. (2018). Theory and practice of data citation. *Journal of the Association for Information Science and Technology*, *69*, 6–20. https://doi.org/10.1002/asi.23917.

Smith, D. R. (2016). Will Publons popularize the scientific peer-review process? *Bioscience*, *66*(4), 265–266. https://doi.org/10.1093/biosci/biw010.

Smith, M. (2014). Data governance: Where technology and policy collide. In J. Ray (Ed.), *Research data management: Practical strategies for information professionals* (pp. 45–59). West Lafayette, IN: Purdue University Press.

Špiranec, S., & Kos, D. (2017). Data literacy and research data management: The Croatian state of affairs. In *European conference on information literacy* (pp. 148–157). Cham: Springer. https://doi.org/10.1007/978-3-319-74334-9_16.

Steinerová, J. (2010). Ecological dimensions of information literacy. *Information Research: An International Electronic Journal*, *15*(4). paper colis719 http://informationr.net/ir/15-4/colis719.html.

Stodden, V., Bailey, D. H., Borwein, R. J., LeVeque, W., & Stein, W. (2013). Setting the default to reproducible computational science research. *SIAM News*, *46*(5), 4–6.

Stvilia, B., Wu, S., & Lee, D. J. (2018). Researchers' uses of and disincentives for sharing their research identity information in research information management systems. *Journal of the Association for Information Science and Technology*, *69*(8), 1035–1045. https://doi.org/10.1002/asi.24019.

Thelwall, M. (2020). The pros and cons of the use of altmetrics in research assessment. *Scholarly Assessment Reports*, *2*(1), 2. https://doi.org/10.29024/sar.10.

Thelwall, M., & Kousha, K. (2015). Web indicators for research evaluation. Part 2: Social media metrics. *El Profesional de La Información*, *24*(5), 607–620. https://doi.org/10.3145/epi.2015.sep.09.

Todorova, T., Garvanova, M., Peteva, M., & Avramova, V. (2019). Comparative findings from data literacy survey in three Bulgarian universities. In *EDULEARN19 proceedings 11th international conference on education and new learning technologies* (pp. 932–940). Valencia: IATED Academy. https://doi.org/10.1177/0340035219849614.

US Congress. (2013). *S.350—Fair Access to Science and Technology Research Act of 2013*. Retrieved from https://www.congress.gov/bill/113th-congress/senate-bill/350.

Van Tuyl, S., & Whitmire, A. L. (2016). Water, water, everywhere: Defining and assessing data sharing in academia. *PLoS One*, *11*(2), e0147942. https://doi.org/10.1371/journal.pone.0147942.

Vilar, P., & Zabukovec, V. (2019). Research data management and research data literacy in Slovenian science. *Journal of Documentation*, *75*(1), 24–43. https://doi.org/10.1108/JD-03-2018-0042.

Virkus, S., Mandre, S., Pals, E., Kurbanoğlu, S., Špiranec, S., Grassian, E., … Roy, L. (2018). Information overload in a disciplinary context. In *European conference on information literacy* (pp. 615–624). Cham: Springer.

Wilkinson, M. D., Dumontier, M., Aalbersberg, I. J., Appleton, G., Axton, M., Baak, A., … Mons, B. (2016). The FAIR Guiding Principles for scientific data management and stewardship. *Scientific Data*, *3*(1), 1–9. https://doi.org/10.1038/sdata.2016.18.

Williams, A. E. (2017). Kudos: Bringing your publications to life? *Information and Learning Science*, *118*(3/4), 114–119. https://doi.org/10.1108/ILS-01-2017-0005.

Woods, D. D., Patterson, E. S., & Roth, E. M. (2002). Can we ever escape from data overload? A cognitive systems diagnosis. *Cognition, Technology & Work*, *4*(1), 22–36. https://doi.org/10.1007/s101110200002.

Yoon, A., & Kim, Y. (2017). Social scientists' data reuse behaviors: Exploring the roles of attitudinal beliefs, attitudes, norms, and data repositories. *Library and Information Science Research*, *39*(3), 224–233. https://doi.org/10.1016/j.lisr.2017.07.008.

Yoon, A., & Lee, Y. Y. (2019). Factors of trust in data reuse. *Online Information Review*, *43*(7), 1245–1262. https://doi.org/10.1108/OIR-01-2019-0014.

Yu, M. C., Wu, Y. C. J., Alhalabi, W., Kao, H. Y., & Wu, W. H. (2016). ResearchGate: An effective altmetric indicator for active researchers? *Computers in Human Behavior*, *55*, 1001–1006. https://doi.org/10.1016/j.chb.2015.11.007.

Zilinski, L. D., Gobens, A., & Briney, K. (2015). University data policies and library data services: Who owns your data? *Bulletin of the Association for Information Science and Technology*, *41*(6), 32–34. https://doi.org/10.1002/bult.2015.1720410613.

Zilinski, L. D., & Sapp Nelson, M. (2014). Thinking critically about data consumption: Creating the data credibility checklist. *Proceedings of the American Society for Information Science and Technology*, *51*(1), 1–4. https://doi.org/10.1002/meet.2014.14505101107.

Further reading

Mons, M. D., Dumontier, M., Aalbersberg, I. J., Appleton, G., Axton, M., Baak, A., … Mons, B. (2016). The FAIR guiding principles for scientific data management and stewardship. *Scientific Data*, *3*(1), 1–9. https://doi.org/10.1038/sdata.2016.18.

RDA FAIR Data Maturity Model Working Group. (2020). *FAIR data maturity model: Specification and guidelines*. https://doi.org/10.15497/rda00050.

Sayogo, D. S., & Pardo, T. A. (2013). Exploring the determinants of scientific data sharing: Understanding the motivation to publish research data. *Government Information Quarterly*, *30*, S19–S31. https://doi.org/10.1016/j.giq.2012.06.011.

Sayre, F., & Riegelman, A. (2019). Replicable services for reproducible research: A model for academic libraries. *College & Research Libraries*, *80*(2), 260–272. https://doi.org/10.5860/crl.80.2.260.

CHAPTER

3

Data quality, the essential "ingredient"

Introduction

In this chapter, the most important issues of research data quality are covered, paying particular attention to attributes and related governance. Quality has been singled out and presented in this separate chapter because it is an essential "ingredient" of both data literacy and research data management (RDM). Caring for the availability of data is becoming obvious for academic libraries and data librarians, but being knowledgeable in the assessment of the authenticity, integrity, and accuracy of digital objects over time should also be a basic asset (Madrid, 2013). This means that data-related services provided by academic libraries, such as quality auditing and verification for research communities, could play a critical role and lead libraries to become data quality hubs in their institutions (Giarlo, 2013).

Quality in any environment is a complex problem in itself. However, it becomes more challenging when we look at it in the light of changing views on the nature of data and information, as described in Chapter 1, which introduced the concepts of data and information. Nonetheless, whatever views we share on their relationship, we perceive data to be a primary intellectual resource that can be subjected to quality assessment (Heidorn, 2011). As the Royal Society (2012) put it, effective data ecology requires all stakeholders of scholarly research not only to elaborate and implement transparent policies and pay attention to custodianship, but also to give special attention to data quality.

A growing number of research funding organizations all over the world have begun to promote the sharing of scientific data. However, merely releasing or posting data is not enough. To achieve the real benefits of sharing, research data needs to be reused by others (Borgman, 2015) (see also Chapter 2 on ecologies). This requirement involves being concerned with

https://doi.org/10.1016/B978-0-12-824475-3.00004-7

data quality. However, the appraisal of data requires deep disciplinary knowledge. Manually evaluating datasets is not only time consuming but expensive, and the automated methods of assessing data are still in their infancy (Ramírez, 2011, p. 21).

Quality assurance is one of the main components of research data management, offered mainly by academic libraries (Kim, 2016). As a broad concept, RDM includes processes undertaken to create organized, documented, accessible, and reusable quality research data (Corti, Van den Eynden, Bishop, & Woollard, 2019). More details are provided about this in Chapter 4.

Data quality and research data management are also connected to data literacy, which begins to play a role as soon as we collect data and carry out a critical analysis of other's interpretation accuracy and correct representation of conclusions. Data literacy is described in more detail in Chapter 5, *Literacies in the age of convergence*.

We can accept that the protectionist strand of media literacy (Hobbs, 2006) shares similarities with information literacy by "applying critical evaluation to ascertain information quality, but extends such analyses to a broader range of media artefacts" (Nichols & Stornaiuolo, 2019, p. 16). Today we can say with certainty that this broadening range encompasses data quality as well. This comprehensive view may accommodate data citations used as an implicit measurement of data quality (Candela, Castelli, Manghi, & Tani, 2015). This chapter therefore aims to cover the most important issues of research data quality, giving particular attention to its attributes and related governance.

The nature of research data quality

The quantity of literature on data quality is overwhelming. This fact is reflected in several reviews that display general data quality issues (Abraham, Schneider, & vom Brocke, 2019; Janssen, van der Voort, & Wahyudi, 2017), as well as point toward the narrower issues related to big data, cloud technologies, and data governance (Al-Ruithe, Benkhelifa, & Hameed, 2018, 2019).

Data used in the corporate world are usually seen as a company asset, which is defined as a resource controlled by a business entity as a result of past events or transactions, and from which "future economic benefits are expected to flow to the entity" (IASB, 2015). Nonetheless, when focusing on the characteristics of research data quality, we must pay attention to the corporate world's considerable interest in the quality of data. Thus some business-related publications on these issues cannot be neglected, even if there are varied differences between values and approaches to data in corporate and academic (research) environments.

We can clearly tell the differences between these two types of data. In academia, research data is "the output from any systematic investigation involving a process of observation, experiment or the testing of a hypothesis, which when assembled in context and interpreted expertly will produce new knowledge" (Pryor, 2012, p. 3).

Research data is produced in the natural sciences, social sciences, and the arts and humanities, independently of the concrete discipline. Research data consists of "heterogeneous objects and items used and contextualized depending on the academic discipline of origin" (Semeler, Pinto, & Rozados, 2017, p. 773).

In Chapter 1, we mentioned that data is one of the cornerstones of the fourth paradigm of scientific research. This widely shared view should be accompanied by beliefs that data quality is a vital element of this ecosystem. As pointed out by Foster, McLeod, Nolin, and Greifeneder (2018), there are general uncertainties related to data. Nonetheless, quality remains a key risk, reflected by the multidimensional concept of data quality, which pays attention to contextual and context-independent properties of individual data values (Nahm & Hammond, 2013). Legal and regulatory requirements also have an impact on data quality (Abraham et al., 2019). The differing nature of big data (Janssen et al., 2017) and cloud computing (Al-Ruithe et al., 2018) also influences data quality. Neither should we forget about the general characteristic of research data that it is based to a certain extent on both predictive and evaluative judgments (Rieh, 2002).

Actors and stakeholders

Taking the value and risks of processing research data into account has different levels. The highest level of the stakeholders, i.e., that of funders, is present on the macro (international, national, sectoral) level. The *meso* level is occupied by organizations and institutions, while researchers and support staff members are at the micro (individual) level (Neylon et al., 2019).

The leading stakeholders of research data quality are undoubtedly the researchers themselves, whose knowledge and skills should exceed the ease and pace of collecting datasets (Whitmire, Boock, & Sutton, 2015). Nonetheless, we can identify a diversity of other relevant stakeholders as well. These stakeholders can fulfill a number of functions and represent several values when supporting the evaluation and maintenance of quality.

Varied research institutions can have an active role in maintaining data quality by paying attention to the technical and scientific quality of research data, to be described later. Several repositories and data centers have developed quality assurance measures and offer a range of services to evaluate the technical quality of datasets (RECODE, 2014). With their expertise, professionals working in these centers are involved in data curation, but can also play an important role in enhancing data management skills of research library staff and data quality. Data curation is increasingly focusing on improving data quality, in order to guarantee authentication of digital resources throughout use and reuse (Sans & Night, 2019).

Trustworthy digital repositories' mission is "to provide reliable, long-term access to managed digital resources to its designated community, now and in the future" (RLG-OCLC, 2002, p. 5). Claims about the trustworthiness of a digital repository should be verifiable, and such decisions can be based on attributes and capabilities that the repository possesses.

With the aim of securing the reusability and long-term preservation of research data, research funders and publishers are urging researchers to deposit their data in certified and accredited repositories. In this context, accreditation or certification in accordance with appropriate standards of repositories is a key to ensuring data quality. Well-founded decisions can be aided by services such as that offered by the Data Seal of Approval (DSA). This service has been one of the most widely used standards for trusted digital repositories (Donaldson, Dillo, Downs, & Ramdeen, 2017). In 2017, it was replaced by the CoreTrustSeal (https://www.coretrustseal.org/) core-level certification that fulfills the same functions (L'Hours, Kleemola, & de Leeuw, 2019).

Scholarly publishers can help to increase the level of data quality by promoting transparency and openness in the following key areas:

- Understanding researchers' problems and motivations, and raising awareness of actual issues;
- Encouraging behavioral and cultural change, and improving the quality and objectivity of the peer-review process;
- Improving scholarly communication infrastructure with journals that publish all scientifically sound research;
- Promoting the registration of scholarly investigations, partnering with data repositories, and providing services that improve data sharing and data curation;
- Increasing incentives for practicing open research with data journals and software journals;
- Implementing data citation and badges for transparency and making research communication more open and accessible, with open-access publishing options (Hrynaszkiewicz, 2019, p. 320).

Many of these efforts demonstrate that scholarly publishers want to exploit the possibilities of research data by providing new products, such as publishing data journals and offering services to enhance data quality (RECODE, 2014). The purpose of data journals goes beyond supporting quality management.

The fate of data quality depends considerably on the successful interventions of data professionals. Although it is not easy to distinguish between the different groups because of their overlapping functions, we will concentrate on three of them: data managers, data librarians, and data scientists. Nonetheless, all these professionals may be involved in quality control.

Data quality attributes

The dimensions of data quality to be considered are various and depend on the objective of the analysis (Daraio et al., 2016). It is therefore not surprising that the literature on data quality is replete with numerous and varied attributes that are often grouped differently.

There are varied perspectives in exhibiting data properties (Laranjeiro, Soydemir, & Bernardino, 2015); therefore we need to handle a considerable number of attributes. For example, in the initial phase of their investigation, Wang and Strong (1996) found no less than 179 data quality attributes. The sometimes frequent repetitions and often overlapping classifications may lead to duplication or triplication of terms or their meanings. However, if we want to paint a broad picture, the portrayal of data quality will be deliberately replete with overlaps.

Despite this situation, there is a need for briefly delineating the relationship between properties, attributes, and characteristics. As explained by Wand and Wang (1996), the properties of things are represented as attributes, which are characteristics assigned to things by humans, depending on purpose and experience. On the other hand, the word *characteristic* is used as a synonym of attribute and denotes a feature of a thing that identifies it.

Such patterns can be seen if we examine the intrinsic and extrinsic nature of data quality. The reason for this is that data quality is defined to a certain extent by the idea that the value of data is based on its fitness for use, which is an extrinsic property (Altman, Marciano, Lee, & Bowden, 2012). In contrast to this, digitized objects "hold intrinsic value, regardless of whether they are ever utilized for any purpose" (Sposito, 2017, p. 2).

The emphasis on intrinsic data quality also appears in the categorization of Strong, Lee, and Wang (1997), who group data quality dimensions into four categories.

Besides accuracy, which is the degree of adherence to the mathematical representation but which also involves correctness and consistency (Daraio et al., 2016), intrinsic data quality includes accuracy, objectivity, believability, and reputation.

Accessibility and access security also pertain to data quality. Contextual attributes of data quality are relevancy, value-added nature, timeliness, the amount of data, and completeness, which—according to Daraio et al. (2016)—is the degree of coverage of knowledge on the phenomena to be represented. Representational data quality falls into the characteristics of interpretability, ease of understanding, concise representation, and consistent representation. Among these attributes, timeliness is understood as the extent to which the age of data is appropriate for the given task (Wand & Wang, 1996).

These four types of data quality are depicted in Table 1.

From the preceding attributes, the degree to which data is seen as credible and true, i.e., believable, appears in a grouping based on four aspects of semiotic representation (empirics, syntactics, semantics, and pragmatics) (Baškarada & Koronios, 2013), displayed in Table 2.

TABLE 1 Data quality attributes.

Intrinsic DQ			
Accuracy	Believability	Objectivity	Reputation
Accessibility DQ			
Accessibility	Access security		
Contextual DQ			
Completeness	Value-addedness	Timeliness	Relevancy
Representational DQ			
Concise representation	Consistent representation	Ease of understanding	Interpretability

TABLE 2 Semiotic attributes of data quality.

Empirics			
Accessibility	Security	Timeliness	Traceability
Syntactics			
Accuracy	Concise representation	Consistency	Ease of manipulation
Semantics			
Ambiguity	Interpretability	Understandability	
Pragmatics			
Added value	Completeness	Relevance	Reputation

From Koltay, T. (2020). Quality of open research data: Values, convergences and governance. Information, 11(4), 175. https://doi.org/10.3390/info11040175.

Taking the semiotic representation gives us a somewhat independent perspective that allows looking at academic or corporate views on quality and dissolving the dichotomy between them. If we take the example of sentences, we can say that their quality is high because they are characterized by correct letters, spelling, and grammar. However, we can perceive poor quality when a nonnative speaker does not understand a given sentence, either partially or in its entirety. This implies that data quality problems are caused not only by engineering flaws (e.g., because of using improper software and/or hardware), but by human factors (brought about, for example, by systematic spelling mistakes). This dictates that data quality should be assessed against predefined specifications for encoding, storage, and communication of data (Baškarada & Koronios, 2013).

According to Wang and Strong (1996), concise representation, appearing previously as part of both intrinsic and semiotic types of quality, can be understood as *compactness*. Nonetheless, its dual role shows that it can be recognized as meaningful on its own account, demonstrating the complexity of approaches to data quality.

The same authors also enumerate the following attributes:

- Appropriate amount, meaning a suitable quantity of the data.
- Believability, which is the degree to which data is seen as credible and true.
- Completeness, which refers to the scope of the information in the data.
- Concise representation, understood as compactness.
- Consistency, meaning that data is continuously presented in the same format.
- Ease of understanding, involving being clear, readable, or understandable.
- Objectivity, meaning the degree of being impartial.
- Relevancy, involving the degree of being applicable, or interesting.
- Reputation, indicating that the data contents or source are held in high consideration.
- Timeliness, denoting the age of the data (Wang & Strong, 1996).

To completeness it can be added that Wand and Wang (1996) see it as the ability of an information system to represent every meaningful state of a real-world system, as well as the extent to which data is of sufficient breadth, depth, and scope.

The last issue, i.e., timeliness, will be addressed in "data governance" section.

There are two more attributes, mentioned by Daraio et al. (2016), that fit into this list:

- Value-added nature that provides a competitive advantage.
- Interpretability, standing for the extent to which the data meaning is explained.

Currency, timeliness, and volatility are time-related, generic dimensions, which are valid in multiple contexts (Laranjeiro et al., 2015). At the same time—if we make arrangements for completeness, comparability, and comprehensiveness—they are key factors in eliciting use (Canares, 2019). Originality, answering the question as to whether the data is genuine or derived from other sources, is of utmost importance, especially when combined with timeliness (Daraio et al., 2016).

Although usually not ascribed as a quality attribute, retrievability of data deserves attention, first of all because data retrieval is a less obvious task than retrieving research papers, which are available in a predominantly textual form (Bugaje & Chowdhury, 2017).

Trust: A distinguished data attribute

Trust (trustworthiness) is one of the most notable data quality attributes. Similarly to many other attributes, it is complex in itself, being influenced by the given subject discipline, the data creators' reputation, and the biases of those who evaluate the data (Daraio et al., 2016). Hence it is not surprising that it comprises several components. Trust is related to and is of extraordinary importance for both research data in general and data stored in repositories (Wolski, Howard, & Richardson, 2017).

Giarlo (2013) underlines that evaluating the extent to which data is trusted depends on a set of subjective factors, such as judging the data to be authentic, seeing its use as acceptable, acknowledging the reputation of those responsible for the creation of the data, and recognizing the biases of the person who is evaluating the data.

Authenticity is based on judgments about the reliability of the instruments used, "the soundness of the underlying theoretical frameworks, the completeness, accuracy, and validity of the data, and ontological consistency within the data" (p. 4). Evaluating authenticity is feasible if the data are understandable by having sufficient context in the form of documentation, metadata, and information on its provenance. Usable data is discoverable and accessible in an appropriate file format. It must allow the judging of its quality with a proper tool to access the data, and the data must be of sufficient integrity to be rendered.

Integrity of data is required for usability, understandability, and authenticity. It assumes that the data turns out identical, at the bit level, to some prior accepted or verified state. Data integrity thus affects overall quality. Trust in the data can be approached by asking if the given dataset has been reviewed by anyone other than its creator (Smith, 2011).

The grade of understandability of data depends on the presence and the appropriate quantity of context in the form of documentation, metadata, or information on provenance. The lack of sufficient context hampers the understanding of data.

The next attribute, usability, is not only a component of trust, but is also an enabling factor, making data discoverable, accessible, and available in a usable format. It may prove data identical at the bit level to some prior accepted or verified state that enables integrity, which is required for understandability and authenticity (Giarlo, 2013).

The elements of trust also include the lineage, version, and error rate of data by being understood and acceptable (Buckland, 2011).

The framework devised for the analysis of trust by Rousseau, Sitkin, Burt, and Camerer (1998) consists of three dimensions. The first one is the propensity of individuals to trust. The second covers trustworthy behavior and is based on the evaluation of people's ability, integrity, and benevolence. The third dimension is defined by levels of trust development achieved.

As Selnes (1998) put it, trust involves competence and communication, while communication, commitment, and conflict handling drive satisfaction. Culture is one of the possible additional critical elements not only in an open science environment but also in accommodating an increasingly diverse range of stakeholders.

According to Wolski et al. (2017), building trust through perceptions about competence is possible by providing evidence that the service provider is able to do what they promise. In the case of research institutions and universities, a benevolent level of trust is inbuilt because of the inherent mission of these institutions. Beside of competence, we also have to count with the need for transparency and openness, which are key attributes of Open Science. There should be information about the objectives of governance, as well as its structure, including governance, maturity, and rigour.

The Framework to Assess Empirical Reproducibility in Biomedical Research, elaborated by McIntosh, Juehne, Vitale, et al. (2017), is not strictly a trust framework, but is a good example of discipline-specific standard processes that aim at reusability through identifying practices by addressing concerns about the lack of trust regarding the reliability and verifiability of biomedical research outputs. Although not mentioned in their paper, this reminds us of the ultimate goal of FAIR principles, i.e., enabling the reuse of data. Although, to achieve this, metadata and data should be well-described so that they can be replicated and/or combined in different settings. This is true even if the FAIR principles do not mention trustworthiness, but requires an accurate description of relevant attributes with metadata. These descriptions, however, demand information on provenance, the presence of which undoubtedly advances trustworthiness.

In the current environment, openness and transparency are a critical part of the process in building trust and will need to be demonstrated by the service provider, whereby research environments are characterized by culture, which has to align with the expectations of the various stakeholders, including end-users (Wolski et al., 2017).

In this framework, culture characterizes the organization that provides the given service. Let us add that research institutions and universities have a benevolent level of trust inbuilt because of their inherent mission, especially if they demonstrate openness and transparency. Commitment is also a key characteristic in building trusted relationships, which can be shown through the website of the provider or may have its origin in the lived experience of the user. Commitment can be documented in terms of service and agreements between partners. Conflicts may arise from different perceptions and expectations about goals or levels of services and should be resolved properly. Efficient communication between the service provider and the end-user is critical at all levels.

Risks should be reduced and satisfaction building should be considered in developing a response to the trust framework, remembering that different perceptions of value propositions have to be managed.

Giarlo (2013) argues that trust depends on subjective judgments of authenticity, acceptability, or applicability of the data, and is also influenced by the given subject discipline, the reputation of those responsible for the creation of the data, and the biases of the persons who are evaluating the data. It is also related to cognitive authority, which has two levels. At an operational level, cognitive authority is the extent to which users think that they can trust the information. On a more general level, cognitive authority refers to influences that a user would recognize as proper because the information therein is thought to be credible and worthy of belief (Rieh, 2002). As explained by Yoon (2014), in the case of data repositories, the perception of trust is based largely on a "lack of deception," but trust in data may not be related to trust in repositories. There are two levels of relatedness of trust to cognitive authority. On the operational level, it is the extent to which users think that they can trust the given piece of information or dataset. On a more general level, cognitive authority refers to influences that a user would recognize as proper because the information therein is thought to be credible and worthy of belief (Rieh, 2002).

Authenticity, which is another essential data quality attribute, measures the extent to which the data is judged to represent the proper ways of conducting scientific research, including the reliability of the instruments used to gather the data, the soundness of underlying theoretical frameworks, and the completeness, accuracy, and validity of the data (Giarlo, 2013). Validity is related to other dimensions, like accuracy, timeliness, completeness, and security (Miller, 1996).

In order to evaluate authenticity, the data must be understandable. The presence of sufficient context in the form of documentation and metadata allows the evaluation of understandability. To achieve this, data has to be usable. To make data usable, it has to be discoverable and accessible and be in a usable file format. The individuals judging data quality need to have at their disposal an appropriate tool to access the data, which has to show sufficient integrity to be rendered. As already stated, integrity of data assumes that the data can be proven to be identical, at the bit level, to some previously accepted or verified state. Data integrity is required for usability, understandability, authenticity, and thus overall quality (Giarlo, 2013).

Fig. 1 shows the relationships between elements of trust, where the numbered elements of the figure are the following:

1. Origin (originality),
2. Methods of collecting and processing,
3. Authenticity,
4. Acceptability,
5. Applicability,
6. Understandability,
7. Subject discipline,
8. Reputation of the creator(s),
9. Biases of the evaluators,
10. **User's confidence,**
11. **"Lack of deception,"**
12. **Independent review, and**
13. **Reuse.**

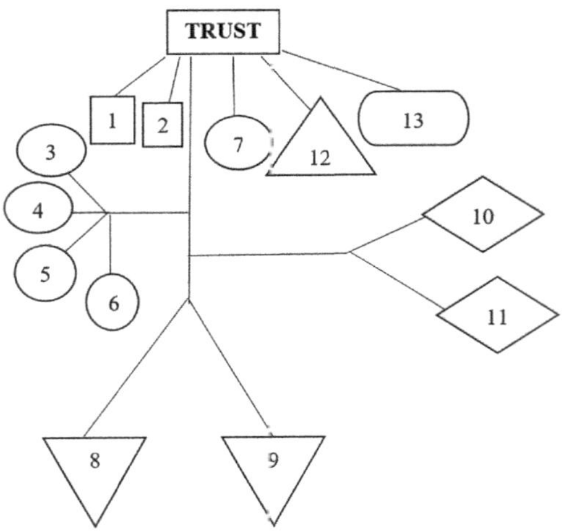

FIG. 1 The relationship between elements of trust. From Koltay, T. (2020). Quality of open research data: Values, convergences and governance. *Information, 11*(4), 175. https://doi.org/10.3390/info11040175.

Some components of trust deserve additional explanation. Authenticity measures the extent to which the data is judged to represent the proper ways of conducting scientific research, including the reliability of the instruments used to gather the data, the soundness of underlying theoretical frameworks, and the completeness, accuracy, and validity of the data. Authenticity, acceptability, and applicability of the data are based on subjective judgments and depend on the given subject discipline (Smith, 2011).

Trust is also a vital element of data governance. As formulated by Sarsfield (2009, p. 38), it "guarantees that data can be trusted and that people can be made accountable for any adverse event that happens because of poor quality."

Technical and scientific quality

On a somewhat different dimension, we can speak about technical quality that involves some aspects, such as completeness and consistency of the dataset, and the use of appropriate standards and software (RECODE, 2014).

As described by the World Data Center for Climate, the minimal set of technical quality criteria consists of the following:

- The number of datasets is correct and bigger than 0,
- The size of every dataset is also bigger than 0,
- The datasets and corresponding metadata are accessible,
- The data sizes are controlled and correct,
- The metadata is consistent with the data,
- The format is correct, and
- Data and its descriptions are consistent (DKZR, 2019).

Technical quality is complemented by scientific quality that is assessed by the research community through pre- and postpublication peer-review. Its assessment is based on the assumption that the peer reviewers do not normally work for some economic benefit, but in order to exercise academic authority, driven by allegiance to the given scientific community (Colepicolo, 2015). We should not forget that—unlike in the case of other scholarly outputs that have established peer-review systems—there is no similar process of validation for data (Yoon, 2017).

Other quality factors

Considering ethical questions of collecting and managing data is a must (Dodds & Wells, 2019). See more details in Chapter 2, "Information and Data Ecologies."

Providing metadata—already mentioned several times throughout this chapter—is indispensable, as it provides information about a variety of quality-related parameters of data, including, e.g., the verification of provenance and authenticity (Royal Society, 2017).

Depending on the type of given data standard and quality attributes considered, data standards may or may not impact data quality. There are at least two worldviews on approaching quality. One is conformance to specifications, and the other is fitness for use. While these are not necessarily different, in practice, there may be divergence between them.

Notwithstanding, the data quality model based on the ISO/IEC 25012:2008 (n.d.) standard should also be mentioned.

When defining a general data quality model, most characteristics are seen in specific contexts of use. The quality characteristics described in this standard are classified into three main categories:

- inherent data quality,
- system-dependent data quality,
- inherent and system-dependent data quality.

The first one refers to data itself in terms of data domain values and possible restrictions, relationships of data values, and metadata.

The second category depends on the technological domain in which data are used and is the degree to which data quality is reached and preserved within a computer system, while the third one is related to both categories.

Table 3 shows the distribution of the 15 data quality categories, pertaining to the previously defined categories.

From the group of inherent categories, accuracy has two separate aspects that represent the correctness of data in regard to syntactics and to semantics, respectively (Syntactic Accuracy and Semantic Accuracy). Completeness signifies that data has values for all expected attributes and related entity instances. Consistency means being coherent with other data and free from contradiction. Credibility signifies being regarded as true and believable, and includes the concept of authenticity (the truthfulness of origins, attributions, and commitments). Currentness is the degree to which data has attributes that are of the right age.

The category of inherent data quality appears in other classifications such as intrinsic data quality, described by Strong et al. (1997), consisting of accuracy, objectivity, believability, and reputation. Mentioned in this chapter several times, accuracy is apparently seen by many authors as one of the decisive factors, even if surrounded by different quality categories in its appearances.

Coming back to the ISO model, the first of the three system-dependent aspects is availability, which is based on attributes that enable data to be retrieved by authorized users and/or applications. Portability enables data to be installed, replaced, or moved from one system to another, preserving

TABLE 3 Inherent and system-dependent data quality.

Inherent DQ	System-dependent DQ	Inherent + system-dependent DQ
Accuracy	Availability	Accessibility
Completeness	Portability	Compliance
Consistency	Recoverability	Confidentiality
Credibility		Efficiency
		Precision
		Traceability
		Understandability

From Koltay, T. (2020). Quality of open research data: Values, convergences and governance. Information, 11(4), 175. https://doi.org/10.3390/info11040175.

its existing quality. Recoverability enables data to maintain and preserve a specified level of operations and quality, even in the event of failure.

From the group, featuring both inherent and system-dependent components, accessibility enables data to be accessed in varied, specific contexts of use. Compliance means adherence to standards, conventions, or regulations relating to data quality. Together with security, availability, and integrity, confidentiality ensures that data is only accessible and interpretable by authorized users. Efficiency signifies the ability of data to be processed and to provide the expected levels of performance. Precision measures the exactness of data attributes. Traceability is the degree to which data has attributes that provide an audit trail of access to the data. Understandability enables data to be read and interpreted by users. Some information about data understandability is usually provided by metadata.

Big data quality

As stated earlier, Volume, Velocity, and Variety are traditionally regarded as main attributes of big data, which can be supplemented by Value and Veracity. The latter attribute is of particular importance to data (and big data) quality by including two aspects: data consistency (or certainty) and data trustworthiness. Not only this attribute but most quality attributes generally applied to research data are apparently valid for big data as well. Nonetheless, there is uncertainty around the understanding of what the data actually means and in what context data is collected, so decision-making quality may be jeopardized (Janssen et al., 2017).

Consistency of big data can be defined by its statistical reliability, while trustworthiness is defined by data origin and collection and processing methods, including trusted infrastructure and facilities. Big data veracity ensures that data are trusted, both on the provenance and the storage side of their lifecycle. Veracity also means that data is authentic and protected from unauthorized access and modification. It also requires:

- Integrity of data,
- Identification of both data and source,
- Trustworthiness of computers and storage platforms,
- Availability and timeliness,
- Accountability and reputation (Demchenko, Grosso, De Laat, & Membrey, 2013).

Managing Variety is a considerable difficulty, because individual data needs to be structured and correlated, regardless of source, and we face the challenges of classifying, prioritizing, and deciding how to notice relationships to Value. Velocity is an important issue because data does not flow on the Internet intermittently but continuously. To gain relevant knowledge, big data requires fast processing, done preferably in real time (Szűts, 2018). Fig. 2 below shows the relationship between veracity and trust.

Floridi (2014) argues the real epistemological challenge of the big data era is that of small patterns. This implies that knowing how to ask and answer questions is crucial; therefore seeing data quality in terms of fit-for-purpose is correct, but this needs to be complemented by a methodology of abstraction, which allows differentiating between quality for producing and consuming data in order to analyze a variety of dimensions, such as accuracy, completeness, consistency, and timeliness.

To predict data quality in large unstructured datasets, it is mostly necessary to analyze the entire dataset first of all, because there is unpredictability of "correct" or "expected" data values. This fact also hampers data optimization at the same time as the data collection (Dresp-Langley et al., 2019).

The incomplete and often uncertain nature of big data negatively influences its quality, thus could become an increasing risk. On the other hand, future research should determine what metrics of quality are adequate to it and how accurate big data needs to be (Abraham et al., 2019). Typically, vast amounts of heterogeneous (big) data are integrated by means of data lake architectures, i.e., systems or repositories of data, stored in its original, natural, or raw format. Data lakes often provide low-quality contents from various data sources, and thus they need to be cleaned (Ceravolo et al., 2018).

On a more general level, Frické (2015, p. 660) states that "science needs more theories and less data," implying what has been expressly aimed at

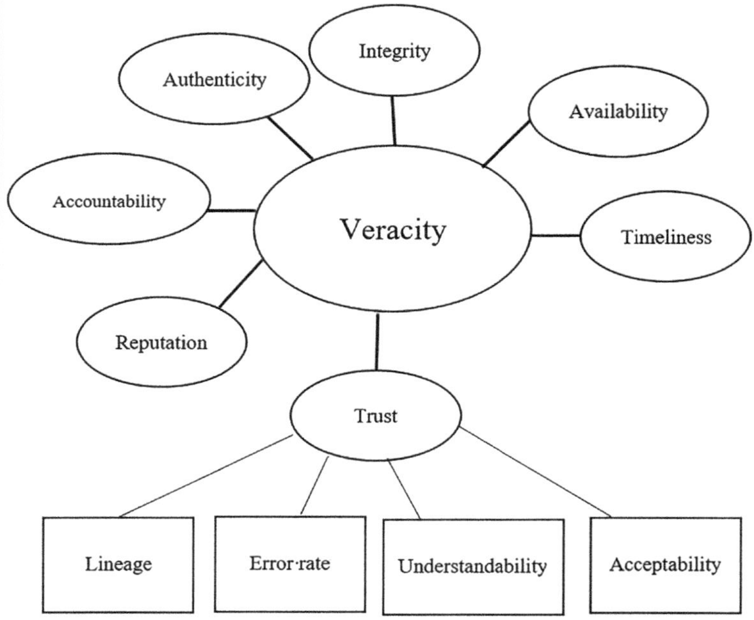

FIG. 2 The relationship between veracity and trust. From Koltay, T. (2020). Quality of open research data: Values, convergences and governance. *Information, 11*(4), 175. https://doi.org/10.3390/info11040175.

by Halme, Komonen, and Huitu (2012), i.e., focusing on quality can be more useful than allowing quantity to dominate.

One of the reasons to think so is that, in big data environments, data processing is not performed on clean datasets from well-known and limited sources. While this is allowed to obtain accurate results, big data may originate from many different and not always consistent and verifiable sources (Hashem et al., 2015).

Last but not least, due to the distinct challenges offered by big data, data governance (discussed in the following section) is critical, especially in environments where data integrity may be at risk (Breaking Big, 2014).

Data governance

As Sarsfield (2009) put it, data governance is like an elephant in a dark room, which can be perceived differently depending on where you touch it. If you touch its tail, it feels like a snake, while if you touch one of its legs, it feels like a tree. Therefore cross-functional perspectives on data governance vary, and we must take this variability into consideration in order to relate data governance to data quality and data literacy properly.

The applicability of data governance to research data management illustrates how the expertise gained in the corporate sector can be used for managing and publishing in other sectors, especially in academia (Willoughby, 2019). Namely, decisions made regarding data quality fall into one of the decision domains of governance frameworks (Khatri & Brown, 2010).

As said earlier, several definitions of data governance underline that it is exercising control over the data within corporate environments (Smith, 2007). While this is true, it seems to be clear that in the academic sector, librarianship as well as library and information science should pay attention to it. This need is asserted, although rather implicitly, by DosSantos (2015), who points out that data governance bears resemblance to the work of a data librarian, because data governance should not be an impediment to data but the driving force behind innovation. The decision-making force of data governance involves decision rights and accountabilities about data based on agreed-upon models, describing who can take what actions, when and under what circumstances, using what methods (DGI, 2015). It specifies a cross-functional framework that formalizes data policies by describing standards and procedures and monitoring compliance (Abraham et al., 2019).

It is important to underline that data governance should not be optional, because it runs horizontally by defining the nature of data and its use in the data management process, not forgetting about metadata, unstructured data, registries, taxonomies, and ontologies.

Successful data governance must be built on clear definitions of its objectives. For instance, it must guarantee that data can be trusted and assure that responsibility will be taken by people who can be made accountable for poor quality (Sarsfield, 2009).

The stakeholders of data governance are identical with those of data quality. In the academic context, researchers, research institutions, funders, publishers, and (although indirectly) the public at large are all interested in good data governance. In the case of researchers, a good understanding of data governance easily addresses their fear of lost rights and benefits. A system of rights, rules, and responsibilities that specifies who can do what with data, data governance mobilizes skills, such as dealing with licensing terms and agreements and having knowledge about copyright, which are already possessed by librarians (Krier & Strasser, 2014). Being knowledgeable of data governance's nature is foundational for providing open data (Weber, Palmer, & Chao, 2012), even though it is also one of the most challenging issues of data sharing, addressed earlier (Krier & Strasser, 2014).

To illustrate the importance of appropriate data governance, we can take the case study presented by Soares (2012) about the unfortunate events surrounding the Mars Climate Orbiter. In 1999, a navigation error directed the Orbiter into a trajectory 170 kilometers lower than the intended altitude above Mars, because NASA engineers used English units (pounds) instead of NASA-specified metric units (newtons). This relatively minor mistake resulted in a huge miscalculation in orbital altitude and in the loss of $328 million. With appropriate attention paid to data governance principles, to the actual details, and to mobilizing data literacy skills, this accident could have been avoided.

Managing and governing

Although data governance is a critical component of any data-quality management strategy (Breaking Big, 2014), it should not be confounded with data management (Al-Ruithe et al., 2019). It is much more correct to say that data-quality management is a subfunction of data management, led by data governance (Otto, 2011). Accordingly, data governance is the starting point for managing data. A formal data governance program has to provide answers to questions such as the availability and access possibilities, provenance, meaning, trustworthiness, and other attributes. As a shared responsibility among all constituents of an institution, it also must provide coordinated, cross-functional approaches and facilitate best practices both to prevent the misuse of institutional data assets and encourage more effective use of these same data assets by the institution itself (ECAR, 2015).

Data governance formalizes behavior related to the definition, production, and usage of data, as well as its quality (Seiner, 2014). Defining the data and how to use it are in the domain of data management, while integrating data into the organization and regulating who is responsible for overseeing the administration of data processes pertain to data governance (Smith, 2007).

Data governance comprises the development of a data-quality strategy; the definition of roles, responsibilities, metrics, and management processes; and the continuous measurement of quality levels. The relationship between data governance and a data-quality strategy is based on a reciprocal whole-part relationship that we could also call a mutual dependence. Nonetheless, we have to emphasize that data governance has the potential to improve data quality, because it can increase numerous quality factors (Abraham et al., 2019).

The five main domains of data governance are:

- Clarifying the role of data as an asset,
- Safeguarding data quality,

- Defining and providing metadata,
- Specifying access requirements, and
- Determining the definition, production, retention, and retirement of data (Khatri & Brown, 2010, p. 150).

As mentioned in this chapter, trust is crucial for data quality. Its importance is present not only in varied general discussions of data quality, but is also emphasized in several definitions of data governance. Sarsfield (2009) formulated this as follows: Data governance "guarantees that data can be trusted and that people can be made accountable for any adverse event that happens because of poor quality" (p. 38).

By reducing the proliferation of incorrect or inconsistent data through continuous analysis, robust data quality can be achieved, and there is also the possibility of enabling the successful monitoring of an information governance program (IBM, 2012). By creating risk-mitigating policies, data governance should care for the management of risks that may arise due to nonconformance with information policies or the lack of control (Abraham et al., 2019).

In research settings, governance structures are needed for managing data related to human subjects, because taking care of sensitive information requires not only establishing standards and norms of practice, but fostering culture change toward better data stewardship (Hartter, Ryan, & MacKenzie, 2013). The culture of data governance must take into account the need for information governance, often used interchangeably with data governance, which is a framework that regulates (among others) information quality and is also of interest for researchers (Lajara & Maçada, 2013).

The place of data governance within the organization

Data governance can refer to organizational bodies, rules, policies, standards, decision rights, accountabilities, and methods of enforcement. It enables better decision-making and protects stakeholders' needs. Let us repeat that there are stakeholders who require interventions for the same or similar issues. It reduces operational friction and encourages adopting common approaches to data issues. It also facilitates building repeatable, standard processes by increasing effectiveness through coordination of efforts and enabling transparency of processes. It is governed by the principles of integrity, transparency, and auditability (DGI, 2015).

Data governance defines who is entitled to make decisions in a given organization and who is held accountable for the decision-making related to data assets (Khatri & Brown, 2010). Governance delineates decision domains, i.e., what decisions must be made to ensure effective management and use of the organization's assets. It also defines the locus of accountability for decision-making, i.e., who makes the decisions (Weill & Ross, 2004). In a word, data governance—by its overarching nature—is one of the prerequisites of effective and efficient data management that substantiates research data services.

Similarly to quality, correct and efficient governance depends as much on technology as on organizational culture (Koltay, 2020). Nonetheless, there is proximity of data governance to information and technology governance (Telino, Massa, Mota, Gomes, & Moreira, 2020).

Obviously, this does not preclude making data transparent by good governance technology. On the contrary, technology helps to identify areas where performance can be improved (ORACLE, 2015). Notwithstanding, accountabilities for managing data quality should not be exclusively assigned to information technology departments.

To be successful in data governance, organizational issues also deserve attention (Weber, Otto, & Österle, 2009). Neither should data quality be treated only informally and in an ambiguous manner, as was done in earlier times by the preponderance of information technology (Al-Ruithe et al., 2018).

Data governance does not need to be threatening and difficult, even though it requires the administration of the behavior around the management of data (Seiner, 2014).

Data governance in big data environments and cloud computing

The traditional principles of data governance apply to big data environments as well. Among the big data types, web data and social media data, as well as machine-to-machine data, deserve special attention. Big data governance is especially important in regard to the acceptable use of data and because underestimating data governance is one of the most common data integration mistakes in environments where big data plays a substantial role (ORACLE, 2015). In the case of big data master data, data governance's focus is less frequently on privacy (Abraham et al., 2019). Data integration can be defined as the combination of processes used to blend data from disparate sources into meaningful and valuable information. It helps to understand, cleanse, monitor, transform, and deliver data, and thus it delivers trusted data from a variety of sources. Data integration solves the problem related to combining data of different provenance by presenting a unified view of these data (Lenzerini, 2002). Although big data integration differs from traditional data integration in many factors (Dong & Srivastava, 2013), it demonstrates the complexity and importance of data governance.

In big data environments, data content changes quickly (Cai & Zhu, 2015); therefore timeliness, defined as the time delay within which meaningful analysis can take place, plays an important role by requiring data to be available, from data generation and acquisition to utilization, as soon as possible (Kim & Cho, 2017). In other words, data must be prepared in a timely manner, without forgetting about the need for consistency and reliability of the data, the trust in its source, and the meaningfulness of the results (Al-Badi, Tarhini, & Khan, 2018).

The nature of data governance may be influenced by the growing importance of cloud computing. The most important differences between the two types is that all traditional data governance policies are handled in-house, while in the case of cloud-based governance it is up to a third party to ensure that guidelines are followed. On-site infrastructure characterizes the former, contrasted to the multisite operation of the latter. This latter is characterized by a loss of control as data governance policies are implemented (Al-Ruithe et al., 2019).

Standards, metadata, stewardship, and data governance

Data governance, stewardship, and information management require the presence of related communities of practice. Change management processes are also needed, and—last but not least—there have to be rewards for good data governance behavior (Smith, 2007).

Seiner (2014) adds to this that valid data governance may require identifying "people who informally already have a level of accountability for the data they define, produce and use to complete their jobs or functions."

Successful data governance depends not only on provisions related to roles in general, but responsibilities connected with appropriate data standards and managed metadata environments (Smith, 2007). Therefore managing metadata is one of the key quality-related processes of data governance because it enables—among others—documenting the provenance of data that enables securing its quality (ORACLE, 2015).

Data stewardship, including a suitable data management infrastructure and using metadata, is directed toward ensuring data quality. This term is sometimes used interchangeably with data governance (Jansen, van den Berg, van Overveld, & Boiten, 2019). It defines checks and balances between different teams, between those who create and collect information, others who manage it, those who use it, and also those who define and introduce standards and compliance requirements (DGI, 2015). Data stewards take care of data assets that do not belong to the stewards themselves, so they represent the concerns of others and ensure that data-related work is performed according to policies and practices as determined through governance. In contrast, data governance is an overall process that brings together cross-functional teams (including data stewards and/or data governors) to make interdependent rules or to resolve issues and to provide services to data stakeholders (Rosenbaum, 2010). More information about data stewards can be found in Chapter 6, discussing the role of data professionals.

Conclusion

Data quality profoundly influences scholarly research, the leading stakeholders of which are not only the researchers themselves, but a variety of other professionals and institutions whose interests show both diversity and convergence. The seemingly endless list of data-quality attributes, having links to a substantial extent to business studies, shows features of a convergence that is similar to the convergences that are central to Chapter 5 that not only describes literacies but also emphasizes various convergences.

Assessing and improving the quality of data involves reconstructing the state of the given dataset. This is followed by assessing the quality along relevant quality dimensions. Assessment also includes data analysis, in order to reach a complete understanding of the data and related architectural and management rules. There is also a need to take the requirements, dataset users, and administrators into account with regard to quality. Identifying critical areas and modeling processes of producing and updating data are also crucial. Measurements can be based on quantitative and qualitative metrics (Batini, Cappiello, Francalanci, & Maurino, 2009).

References

Abraham, R., Schneider, J., & vom Brocke, J. (2019). Data governance: A conceptual framework, structured review, and research agenda. *International Journal of Information Management*, *49*, 424–438. https://doi.org/10.1016/j.ijinfomgt.2019.07.008.

Al-Badi, A., Tarhini, A., & Khan, A. I. (2018). Exploring big data governance frameworks. *Procedia Computer Science*, *141*, 271–277. https://doi.org/10.1016/j.procs.2018.10.181.

Al-Ruithe, M., Benkhelifa, E., & Hameed, K. (2018). Data governance taxonomy: Cloud versus non-cloud. *Sustainability*, *10*(1), 95. https://doi.org/10.3390/su10010095.

Al-Ruithe, M., Benkhelifa, E., & Hameed, K. (2019). A systematic literature review of data governance and cloud data governance. *Personal and Ubiquitous Computing*, *23*(5/6), 839–859. https://doi.org/10.1007/s00779-017-1104-3.

Altman, M., Marciano, G., Lee, C., & Bowden, H. (2012). Mitigating threats to data quality throughout the curation lifecycle. In *Curating for quality: Ensuring data quality to enable new science* (pp. 1–119). Arlington County, VA: National Science Foundation. https://doi.org/10.1002/asi.23480.

Baškarada, S., & Koronios, A. (2013). Data, information, knowledge, wisdom (DIKW): A semiotic theoretical and empirical exploration of the hierarchy and its quality dimension. *Australasian Journal of Information Systems*, *18*(1), 5–24.

Batini, C., Cappiello, C., Francalanci, C., & Maurino, A. (2009). Methodologies for data quality assessment and improvement. *ACM Computing Surveys*, *41*(3), 1–52. https://doi.org/10.1145/1541880.1541883.

Borgman, C. L. (2015). *Big data, little data, no data: Scholarship in the networked world*. Cambridge, MA: MIT Press.

Breaking Big. (2014). Breaking big: When big data goes bad. In *The importance of data quality management in big data environments*. New, York, NY: Information Builders.

Buckland, M. (2011). Data management as bibliography. *Bulletin of the American Society for Information Science and Technology*, *37*(6), 34–37. https://doi.org/10.1002/bult.2011.1720370611.

Bugaje, M., & Chowdhury, G. (2017). Is data retrieval different from text retrieval? An exploratory study. In S. Choemprayong, F. Crestani, & S. Cunningham (Eds.), *Lecture notes in computer science: Vol. 10647. Digital libraries: Data, information, and knowledge for digital lives. ICADL 2017*. Cham: Springer. https://doi.org/10.1007/978-3-319-70232-28.

Cai, L., & Zhu, Y. (2015). The challenges of data quality and data quality assessment in the big data era. *Data Science Journal, 14*. https://doi.org/10.5334/dsj-2015-002.

Canares, M. (2019). Open data around the world: South, East, and Southeast Asia. In T. Davies, S. Walker, M. Rubinstein, & F. Perini (Eds.), *The state of open data: Histories and horizons* (pp. 535–548). Cape Town and Ottawa: African Minds and International Development Research Centre. http://stateofopendata.od4d.net.

Candela, L., Castelli, D., Manghi, P., & Tani, A. (2015). Data journals: A survey. *Journal of the Association for Information Science and Technology, 66*(9), 1747–1762. https://doi.org/10.1002/asi.23358.

Ceravolo, P., Azzini, A., Angelini, M., Catarci, T., Cudré-Mauroux, P., Damiani, E., … Sattler, K. U. (2018). Big data semantics. *Journal on Data Semantics, 7*(2), 65–85. https://doi.org/10.1007/s13740-018-0086-2.

Colepicolo, E. (2015). Information reliability for academic research: Review and recommendations. *New Library World, 116*(11/12), 646–660. https://doi.org/10.1108/NLW-05-2015-0040.

Corti, L., Van den Eynden, V., Bishop, L., & Woollard, M. (2019). *Managing and sharing research data: A guide to good practice* (p. 2). SAGE Publications Limited.

Daraio, C., Lenzerini, M., Leporelli, C., Naggar, P., Bonaccorsi, A., & Bartolucci, A. (2016). The advantages of an ontology-based data management approach: Openness, interoperability and data quality. *Scientometrics, 108*(1), 441–455. https://doi.org/10.1007/s11192-016-1913-6.

Demchenko, Y., Grosso, P., De Laat, C., & Membrey, P. (2013). Addressing big data issues in scientific data infrastructure. In *2013 international conference on collaboration technologies and systems (CTS)* (pp. 48–55). New York, NY: IEEE.

DGI. (2015). *Definitions of data governance*. Data Governance Institute. Retrieved from http://www.datagovernance.com/adgdatagovernancedefinition.

DKZR. (2019). *Quality assurance of data*. Deutsche Klimarechenzentrum. Retrieved from https://www.dkrz.de/up/services/data-distribution/data-publication/quality-assurance-of-data.

Dodds, L., & Wells, P. (2019). Issues in open data: Data infrastructure. In T. Davies, S. Walker, M. Rubinstein, & F. Perini (Eds.), *The state of open data: Histories and horizons* (pp. 260–273). Cape Town and Ottawa: African Minds and International Development Research Centre. http://stateofopendata.od4d.net.

Donaldson, D. R., Dillo, I., Downs, R., & Ramdeen, S. (2017). The perceived value of acquiring data seals of approval. *International Journal of Digital Curation, 12*(1), 130–151. https://doi.org/10.2218/ijdc.v12i1.481.

Dong, X. L., & Srivastava, D. (2013). Big data integration. In *2013 IEEE 29th international conference on data engineering (ICDE)* (pp. 1245–1248). New York, NY: IEEE. https://doi.org/10.1109/ICDE.2013.6544914.

DosSantos, J. (2015). *What librarians can teach us about managing big data*. InFocus. https://infocus.emc.com/joe_dossantos/what-librarians-can-teach-us-about-managing-big-data.

Dresp-Langley, B., Ekseth, O. K., Fesl, J., Gohshi, S., Kurz, M., & Sehring, H. W. (2019). Occam's razor for big data? On detecting quality in large unstructured datasets. *Applied Sciences, 9*(15), 3065. https://doi.org/10.3390/app9153065.

ECAR. (2015). *The compelling case for data governance*. EDUCAUSE ECAR Working Group. Retrieved from http://www.educause.edu/library/resources/compelling-case-data-governance.

Floridi, L. (2014). Big data and information quality. In L. Floridi, & P. Illari (Eds.), *The philosophy of information quality. Synthese library* (p. 358). Cham: Springer. https://doi.org/10.1007/978-3-319-07121-3_15 303-315.

Foster, J., McLeod, J., Nolin, J., & Greifeneder, E. (2018). Data work in context: Value, risks, and governance. *Journal of the Association for Information Science and Technology, 69*(12), 1414–1427. https://doi.org/10.1002/asi.24105.

Frické, M. (2015). Big data and its epistemology. *Journal of the Association for Information Science and Technology, 66*(4), 651–661. https://doi.org/10.1002/asi.23212.

Giarlo, M. (2013). Academic libraries as quality hubs. *Journal of Librarianship and Scholarly Communication, 1*(3), 1–10. https://doi.org/10.7710/2162-3309.1059.

Halme, P., Komonen, A., & Huitu, O. (2012). Solutions to replace quantity with quality in science. *Trends in Ecology, & Evolution, 27*(11), 586. https://doi.org/10.1016/j.tree.2012.08.007.

Hartter, J., Ryan, S. J., & MacKenzie, C. A. (2013). Spatially explicit data: Stewardship and ethical challenges in science. *PLoS Biology, 11*(9), e1001634. https://doi.org/10.1371/journal.pbio.1001634.

Hashem, I. A. T., Yaqoob, I., Anuar, N. B., Mokhtar, S., Gani, A., & Ullah, K. S. (2015). The rise of "big data" on cloud computing: Review and open research issues. *Information Systems, 47*, 98–115. https://doi.org/10.1016/j.is.2014.07.006.

Heidorn, P. B. (2011). The emerging role of libraries in data curation and e-science. *Journal of Library Administration, 51*(7/8), 662–672. https://doi.org/10.1080/01930826.2011.601269.

Hobbs, R. (2006). Multiple visions of multimedia literacy: Emerging areas of synthesis. In M. McKenna (Ed.), *International handbook of literacy and technology* (pp. 15–26). Mahwah, NJ: Lawrence Erlbaum.

Hrynaszkiewicz, I. (2019). Publishers' responsibilities in promoting data quality and reproducibility. In A. Bespalov, M. Michel, & T. Steckle (Eds.), *Handbook of experimental pharmacology*. Cham: Springer. https://doi.org/10.1007/164_2019_290.

IASB. (2015). *IFRS for SMEs* (p. 15). London: International Accounting Standards Board.

IBM. (2012). *Successful information governance through high-quality data*. Somers, NY: IBM Corporation.

ISO/IEC 25012:2008 (n.d.). *Software engineering. Software product quality requirements and evaluation (SQuaRE)*. Retrieved from https://www.iso.org/standard/35736.html.

Jansen, P., van den Berg, L., van Overveld, P., & Boiten, J. W. (2019). Research data stewardship for healthcare professionals. In P. Kubben, M. Dumontier, & A. Dekker (Eds.), *Fundamentals of clinical data science* (pp. 37–53). Cham: Springer. https://doi.org/10.1007/978-3-319-99713-1_4.

Janssen, M., van der Voort, H., & Wahyudi, A. (2017). Factors influencing big data decision-making quality. *Journal of Business Research, 70*, 338–345. https://doi.org/10.1016/j.jbusres.2016.08.007.

Khatri, V., & Brown, C. V. (2010). Designing data governance. *Communications of the ACM, 53*(1), 148–152. https://doi.org/10.1145/1629175.1629210.

Kim, J. (2016). Who is teaching data: Meeting the demand for data professionals? *Journal of Education for Library and Information Science, 57*(2), 161–173. https://doi.org/10.3138/jelis.57.2.161.

Kim, Y., & Cho, J. (2017). Data governance framework for big data implementation with a case of Korea. In *2017 IEEE international congress on big data, Honolulu, HI* (pp. 384–391). New York, NY: IEEE.

Koltay, T. (2020). Quality of open research data: Values, convergences and governance. *Information, 11*(4), 175. https://doi.org/10.3390/info11040175.

Krier, L., & Strasser, C. A. (2014). *Data management for libraries: a LITA guide*. American Library Association.

Lajara, T. T., & Maçada, A. C. G. (2013). *Information governance framework: The defense manufacturing case study*. https://aisel.aisnet.org/amcis2013/DataQuality/GeneralPresentations/2/.

Laranjeiro, N., Soydemir, S. N., & Bernardino, J. (2015). A survey on data quality: Classifying poor data. In *IEEE 21st Pacific Rim international symposium on dependable computing (PRDC)* (pp. 179–188). New York, NY: IEEE.

Lenzerini, M. (2002). Data integration: A theoretical perspective. In *Proceedings of the twenty-first ACM SIGMOD-SIGACT-SIGART symposium on principles of database systems* (pp. 233–246). https://doi.org/10.1145/543613.543644.

L'Hours, H., Kleemola, M., & de Leeuw, L. (2019). CoreTrustSeal: From academic collaboration to sustainable services. *IASSIST Quarterly*, *43*(1), 1–17. https://doi.org/10.29173/iq936.

Madrid, M. M. (2013). A study of digital curator competences: A survey of experts. *International Information and Library Review*, *45*(3/4), 149–156. https://doi.org/10.1016/j.iilr.2013.09.001.

McIntosh, L. D., Juehne, A., Vitale, C. R. H., et al. (2017). Repeat: A framework to assess empirical reproducibility in biomedical research. *BMC Medical Research Methodology*, *17*, 143. https://doi.org/10.1186/s12874-017-0377-6.

Miller, H. (1996). The multiple dimensions of information quality. *Information Systems Management*, *13*(2), 79–82. https://doi.org/10.1080/10580539608906992.

Nahm, M., & Hammond, W. E. (2013). Data standards ≠ data quality. *Studies in Health Technology and Informatics*, *192*, 1208.

Neylon, C., Belsø, R., Bijsterbosch, M., Cordewener, B., Foncel, J., & Laakso, M. (2019). *Open scholarship and the need for collective action*. Bristol: Knowledge Exchange.

Nichols, T. P., & Stornaiuolo, A. (2019). Assembling "digital literacies": Contingent pasts, possible futures. *Media and Communication*, *7*(2), 14–24. https://doi.org/10.17645/mac.v7i2.1946.

ORACLE. (2015). *The five Most common Big data integration mistakes to avoid*. Redwood Shores, CA: Oracle Corporation.

Otto, B. (2011). Data governance. *Business, & Information Systems Engineering*, *4*, 241–244. https://doi.org/10.1007/s12599-011-0162-8.

Pryor, G. (2012). *Managing research data*. London: Facet Publishing.

Ramírez, M. L. (2011). Whose role is it anyway? A library practitioner's appraisal of the digital data deluge. *Bulletin of the American Society for Information Science and Technology*, *37*(5), 21–23. https://doi.org/10.1002/bult.2011.1720370508.

RECODE. (2014). *RECODE policy recommendations for open access to research data*. RECODE Project Consortium. https://trilateralresearch.co.uk/wp-content/uploads/2018/09/RECODE-D5.1-POLICY-RECOMMENDATIONS-FINAL.pdf.

Rieh, S. Y. (2002). Judgment of information quality and cognitive authority in the web. *Journal of the American Society for Information Science and Technology*, *53*(2), 145–161. https://doi.org/10.1002/asi.10017.

RLG/OCLC Working Group on Digital Archive Attributes. (2002). *Trusted digital repositories: Attributes and responsibilities*. Mountain View, CA: RLG. Retrieved from http://www.oclc.org/research/activities/past/rlg/trustedrep/repositories.pdf.

Rosenbaum, S. (2010). Data governance and stewardship: Designing data stewardship entities and advancing data access. *Health Services Research*, *45*(5), 1442–1455. https://doi.org/10.1111/j.1475-6773.2010.01140.x.

Rousseau, D. M., Sitkin, S. B., Burt, R. S., & Camerer, C. (1998). Not so different after all: A cross-discipline view of trust. *Academy of Management Review*, *23*(3), 393–404.

Royal Society. (2012). *Science as an open Enterprise*. London: Royal Society Science Policy Centre.

Royal Society. (2017). *Data management and use: Governance in the 21st century: A joint report by the British Academy and the Royal Society*. London: British Academy, The Royal Society. https://royalsociety.org/~/media/policy/projects/data-governance/data-managementgovernance.pdf.

Sans, S., & Night, W. S. (2019). Introduction. In *HathiTrust: Large-scale data repository in the humanities*. https://akthom.gitbooks.io/hkdcws/content/Introduction/Introduction.html.

Sarsfield, S. (2009). *The data governance imperative: A business strategy for corporate data*. Ely: IT Governance Publishing.

Seiner, R. S. (2014). *Non-invasive data governance: The path of least resistance and greatest success*. Basking Ridge, NJ: Technics Publications.

Selnes, F. (1998). Antecedents and consequences of trust and satisfaction in buyer-seller relationships. *European Journal of Marketing, 32*(3/4), 305–322. https://doi.org/10.1108/03090569810204580.

Semeler, A. R., Pinto, A. L., & Rozados, H. B. F. (2017). Data science in data librarianship: Core competencies of a data librarian. *Journal of Librarianship and Information Science, 51*(3), 771–780. http://journals.sagepub.com/doi/10.1177/0961000617742465.

Smith, A. M. (2007). Data governance best practices—The beginning. *EIMInsight, 1*(1). Retrieved from http://www.eiminstitute.org/library/eimi-archives/volume-1-issue-1-march-2007-edition/data-governance-best-practices-2013-the-beginning.

Smith, M. (2011). Communicating with data: New roles for scientists, publishers and librarians. *Learned Publishing, 24*(3), 203–205. https://doi.org/10.1087/20110308.

Soares, S. (2012). *Big data governance: An emerging imperative*. Boise, ID: MC Press.

Sposito, F. A. (2017). *What do data curators care about? Data quality, user trust, and the data reuse plan*. Paper presented at IFLA WLIC 2017 http://library.ifla.org/1797/.

Strong, D. M., Lee, Y. W., & Wang, R. Y. (1997). Data quality in context. *Communications of the ACM, 40*(5), 103–110. https://doi.org/10.1145/253769.253804.

Szűts, Z. (2018). *Online: Az internetes kommunikáció és média története, elmélete és jelenségei*. Budapest: Wolters Kluwer (in Hungarian).

Telino, V., Massa, R., Mota, I., Gomes, A., & Moreira, F. (2020). A methodology for creating a macro action plan to improve IT use and its governance in organizations. *Information, 11*(9), 427. https://doi.org/10.3390/info11090427.

Wand, Y., & Wang, R. Y. (1996). Anchoring data quality dimensions in ontological foundations. *Communications of the ACM, 39*(11), 86–95. https://doi.org/10.1145/240455.240479.

Wang, R. Y., & Strong, D. M. (1996). Beyond accuracy: What data quality means to data consumers. *Journal of Management Information Systems, 12*(4), 5–33. https://doi.org/10.1080/07421222.1996.11518099.

Weber, K., Otto, B., & Österle, H. (2009). One size does not fit all—A contingency approach to data governance. *Journal of Data and Information Quality (JDIQ), 1*(1), 4. https://doi.org/10.1145/1515693.1515696.

Weber, N. M., Palmer, C. L., & Chao, T. C. (2012). Current trends and future directions in data curation research and education. *Journal of Web Librarianship, 6*(4), 305–320. https://doi.org/10.1080/19322909.2012.730358.

Weill, P., & Ross, J. W. (2004). *IT governance: How top performers manage IT decision rights for superior results*. Boston, MA: Harvard Business School Press.

Whitmire, A. L., Boock, M., & Sutton, S. C. (2015). Variability in academic research data management practices: Implications for data services development from a faculty survey. *Program, 49*(4), 382–407. https://doi.org/10.1108/PROG-02-2015-0017.

Willoughby, S. (2019). Open data and the environment. In T. Davies, S. Walker, M. Rubinstein, & F. Perini (Eds.), *The state of open data: Histories and horizons* (pp. 103–118). Cape Town and Ottawa: African Minds and International Development Research Centre. http://stateofopendata.od4d.net.

Wolski, M., Howard, L., & Richardson, J. (2017). A trust framework for online research data services. *Publications, 5*(2), 14. https://doi.org/10.3390/publications5020014.

Yoon, A. (2014). End users' trust in data repositories: Definition and influences on trust development. *Archival Science, 14*, 17–34. https://doi.org/10.1007/s10502-013-9207-8.

Yoon, A. (2017). Data reusers' trust development. *Journal of the Association for Information Science and Technology, 68*(4), 946–956. https://doi.org/10.1002/asi.23730.

Further reading

Cappiello, C., Di Noia, T., Marcu, B. A., & Matera, M. (2016). A quality model for linked data exploration. In A. Bozzon, P. Cudre-Maroux, & C. Pautasso (Eds.), *Lecture notes in computer science: Vol. 9671. Web engineering. ICWE 2016*. Cham: Springer. https://doi.org/10.1007/978-3-319-38791-8_25.

CHAPTER

4

Research data management

Introduction

Activities related to the management of research data were identified as a major trend in academic library work by the Association of College & Research Libraries in 2012 and 2014 (ACRL, 2012, 2014). In 2016 this discussion continued through a focus on professional development for librarians providing research data services (RDS) (ACRL, 2016). Although important, these overviews only offer one perspective.

The concept of research data is more specific than that of data in general. However, as Robinson (2016, p. 74) reminds us, "this is perhaps readily dealt with as 'data collected as part of a research project', in contrast to data collected or digitised for curation and preservation. However, the question immediately arises as to whether data collected for curation and preservation cannot then become research data, and correspondingly, whether at some point, research data is archived for curation and preservation."

Research data is undoubtedly produced and owned by the researchers themselves. Nevertheless, the need for managing and curating research data in order to ensure continuity, transparency, and accountability should be met by the libraries as well (Sesartić, Fischlin, & Töwe, 2016).

Unfortunately, there is no single term used exclusively for the related complex of service activities; we can choose between research data management (RDM) and research data services (RDS), since the similarities and differences between the two are apparently far from being clarified. Considering this, in this book the services related to research data management are called simply RDM, or RDM services. Fortunately, the typology of these latter concepts clearly shows that they are usually divided into informational and technological services. The difficulties in naming and applying RDM may also be tied to the influence of the VUCA (Volatility, Uncertainty, Complexity, and Ambiguity) phenomenon, which

Research Data Management and Data Literacies
https://doi.org/10.1016/B978-0-12-824475-3.00002-3

affects a number of academic library tasks and requires managing change in order to meet the related challenges, quite apparent in the case of RDM (Benedetti et al., 2020).

The nature of research data management

RDM embraces the entire research cycle, with the aim of "making the research process as efficient as possible and to facilitate cooperation with others. More specifically, RDM helps to protect data, it facilitates in sharing the data with others and it ensures that research data is findable, accessible and (re)usable" (TU/e, n.d.).

RDM can also be defined as a "series of practices for dealing with resources," thus continuing some of the expertise and functions that libraries have held and perfected since antiquity, but embracing (digital) technologies (Willaert et al., 2019, p. 5).

RDM services involve the broad stages of "planning and preparing, conducting, archiving, publishing and disseminating, using a research lifecycle model." Some of these activities are repeated throughout this life cycle (Yu, Deuble, & Morgan, 2017, p. 287).

The development of RDM

A relatively early initiative by LIBER (the Association of European Research Libraries) underlined the need for libraries being involved in the processes of supporting research data management. The importance of this initiative was acknowledged by the League of European Research Universities (LERU), because successful RDM requires cooperation between librarians and researchers, especially in clarifying what is expected of researchers when describing and citing data (LERU, 2013). LIBER set up 10 recommendations for libraries to get started in research data management:

- Offer research data management support, related to data management plans and intellectual property rights, among other factors;
- Provide metadata services for research data and engage in the development of metadata and data standards;
- Create data librarian posts and develop professional staff skills for data librarianship;
- Actively participate in the institutional development of research data and open data policies;
- Collaborate and build partnerships with researchers, research groups, data archives, and data centers;
- Provide services for storage, discovery, and permanent access throughout the life cycle of research data;

- Promote research data citation by applying persistent identifiers to research data;
- Provide an institutional Data Catalogue or Data Repository, depending on available infrastructure;
- Get involved in subject-specific data management practice;
- Offer or mediate secure storage for dynamic and static research data in cooperation with institutional IT units and/or seek exploitation of appropriate services.

Services should depend on several factors, including the size of the institution and the vision of its management. However, there is no need for every institution to implement all the recommended components of research data services. Neither is it always feasible to follow the same set of recommendations in different libraries (Christensen-Dalsgaard et al., 2012).

The trends of data-intensive research have influenced some scholars to characterize the new research landscape as Research 2.0, Science 2.0, or e-science (Koltay, 2019a). Research 2.0 requires high-level attention to data management and data curation, data citation, and data quality (Vilar & Zabukovec, 2019). These terms cover a new way of practicing scholarly research, based on technical tools and devices built with contemporary data-intensive technology (Semeler & Pinto, 2020). The growing attention on making research processes transparent (Sayre & Riegelman, 2019), along with many other developments, could not have evolved without certain antecedents.

As Willaert et al. (2019, p. 5) stated, at the Library of Alexandria the challenges of "creating, transferring and preserving knowledge were faced by theoreticians and practitioners alike. Under one roof, gaps were thus bridged between knowledge management strategies." On the other hand, for a long period libraries have been seen as dispensaries of books and articles (Jahnke, Asher, & Keralis, 2012). Today academic libraries are on their way to becoming one of the top players in supporting data-intensive research (ACRL, 2014; NMC, 2014), or at least this is their aspiration.

A number of past paradigms that have influenced the nature of academic librarianship have been described by Martell (2009). One paradigm was the *Ownership* or *Collections* paradigm. It evolved after World War II and reached its zenith in the 1960s. It was built on the assumption that library systems in higher education would be able to collect all documents to adequately satisfy the institutions' scholarly and teaching needs. Such support allowed for a broad range of interpretations, but it proved to be unsustainable. It was then superseded by the *Access* paradigm, which gave more attention to and made use of resource sharing, from the late 1970s until the end of the 20th century. Widespread access to digital material, in particular the availability of digital full text, made ownership in its traditional sense not practical, so the *iAccess* paradigm followed. More recently, the emergence and growing prevalence of social media created an opportunity to add a social dimension to iAccess, forging in this way the *sAccess* paradigm, which takes account of social networking features.

Today, it is often difficult to disentangle the relationship between the features of Research 2.0 induced by social media and research data management, which are now needed for and by the data-intensive forms of research. Deciding to what extent data-intensive research pertains to iAccess and sAccess is a similarly demanding task. Nonetheless, both paradigms have influenced these developments to a certain degree.

Another way of reflecting on the changes in the digital environment is the concept of the *inside-out library*, which posits that libraries are increasingly supporting the creation, curation, and discoverability of institutional products, including research data. In contrast to the outside-in library model that was dominant earlier, this model focuses on sharing resources that are unique for potential users outside the given institution (Dempsey, 2017).

The readiness for providing RDM services

The appearance of data-intensive research prompted academic libraries to become service providers for scholars who work with research data. Although recognizing the need for providing RDM services in a broad sense is an imperative for libraries worldwide, it shows divergent patterns and extent in varied countries and in their academic libraries.

The nature of these differences is not a terminological one, although we can speak about *readiness* (Chawinga & Zinn, 2020; Chiware & Becker, 2018; Koltay, 2016, 2019a) or *maturity* (Cox, Kennan, Lyon, & Pinfield, 2017). To describe the actual experiences related to academic libraries' response to the emerging requirements, the word *preparedness* can also be used (Tang & Hu, 2019).

It is rather obvious that part of the differences might be related to the simple fact that open data appeared much later than open access, awareness of which could work itself out earlier and to a longer extent. There is also a feeling that the spread of open data was not only much slower but also much more uneven, compared to that of the open access movement (Koltay, 2019b).

Perrier, Blondal, and MacDonald (2018, pp. 177–178) put this as follows:

> "Although libraries had already been collecting and managing research data, providing a higher level of support to researchers had not been a regular part of their activities. Reactionary responses were the norm, as sufficient time was not provided to be proactive. A significant difficulty was to understand the researchers' perspective and aligning with this so that meaningful services could be offered."

Tang and Hu (2019) examined respondents' perceptions on both the current challenges and future roles of RDM services. Using an online questionnaire, they conducted an international survey in 2018. From 5 continents, 241 practicing librarians were surveyed. Eighty-seven respondents answered the question about their preparedness, with 70.1% of them feeling prepared. Those who felt unprepared listed most often the lack of training (30.8%) and a deficit in knowledge and skills (26.9%).

In their bibliometric study, Hsiao and Chen (2020) measured varied relationships between different feature words, extracted from 21,066 research articles published in 44 LIS journals. Among the top 25 subfields, *Research Data Management & Data Curation* was found. For the period between 2014 and 2018, they included the words and/or expressions *data, data curation, data management, curator, data citation, data share, data science*, and *data repository* (Hsiao & Chen, 2020). However, *Open Access and related issues* appeared with much higher frequency than *Research Data Management & Data Curation*.

A survey targeting the majority of libraries serving Canada's research-intensive universities (13 out of 15), by Ducas, Michaud-Oystryk, and Speare (2020), revealed that from among general research support activities (information discovery, grant application support, systematic reviews, and bibliometric services) data management had the lowest share, with 22.55%.

Faniel and Connaway (2018) have found that—although many librarians understand the challenging nature of RDM activities—the respondents of their study described data-related work as interesting, enjoyable, satisfying, and rewarding. Other perceived benefits found in this study included learning new things, building relationships with researchers, and being seen as a source of support.

Skills and competencies

According to Schmidt and Shearer (2016), RDM encompasses a wide array of activities across the research data life cycle, "requiring a high level of interaction with researchers, and working with other support services."

They provided the following list of competencies:

1. Providing access to data
   - Identifying and locating datasets,
   - Providing support for access and reuse conditions of data,
   - Identifying software/code in order to understand and reuse data,
   - Giving advice on informatics, discovery, and analytical tools,
   - Providing support for data citation and referencing.
2. Awareness and support for managing data

   Policy and advocacy:
   - Articulating the benefits of research data management,
   - Promoting data sharing and reuse,
   - Providing education about relevant research data policies.

Support and training:
- Raising awareness of the need and options of RDM,
- Teaching data literacy,
- Assisting with completing RDM plans,
- Identifying effective service workflows,
- Identifying data standards and formats, and metadata standards,
- Identifying repositories/data archives for deposit,
- Advising about software management.

Data audits:
- Identifying the range of datasets on campus,
- Developing data curation profiles.

3. Managing a data collection

Data collection:
- Preparing data for deposit,
- Appraising, selecting, and ingesting of data into repository.

Data management:
- Managing data collection and metadata.

Data preservation:
- Ensuring authenticity,
- Refreshing digital media and migrating data,
- Undertaking backup and storage.

In the environment dominated by textual information, librarians have acquired a number of useful skills. Today they still can and should make use of these traditional skills when dealing with data. Meeting this challenge requires all librarians to have a proper conceptual understanding of data, and to acquire the skills and abilities "to find, extract, collect, clean, organize, analyse, and present data," even if they do not serve users who deal with data (Robinson & Bawden, 2017, p. 314). In other words, RDM aims at "ensuring that data is well preserved, accessed and used as long as it remains relevant and of value to the societies" (Chawinga & Zinn, 2020, p. 1).

Research data management is relevant if it improves research economy, the main aim of which is setting up the conditions for improving efficiency by enabling quick and easy access to data, warranting its sustainability, and facilitating a proper return on investment (Hinrichs-Krapels & Grant, 2016). However, there must also be integrity that ensures reliability, honesty, respect, and responsibility of all activities and results associated with the research process. This requires researchers to be accountable for the research process in its entirety by focusing on ensuring and improving the quality of scientific work, and meeting the requirements that have been set before engaging in the research (ALLEA, 2017).

Assessing research integrity (responsible conduct of research) therefore demands:

- Assessing responsible research practices,
- Valuing the completeness of reporting,
- Rewarding the practice of open science,
- Recognizing other essential tasks, like peer review and mentoring (WCRIF, 2019).

In an exploratory study, Zhang and Eichmann-Kalwara (2019) examined Scopus searches, using simple synonyms of the phrase *research data management* and the words most frequently associated with this expression. The broader term *data management* was excluded from the search queries, because it retrieved a high number of hits related to programming, data mining, machine learning, etc.

This search brought 1913 relevant documents from varied types of sources and publication years, containing 23,402 cited references. Research areas (clusters) focusing on various aspects of RDM were identified. The top three clusters included scientific collaboration, research support services, and data literacy. From these, data literacy was the youngest one.

A US survey by Saunders (2020) studying the evolution of research subfields in LIS during two periods (between 2009–13 and 2014–18) showed that 17% of the responding academic librarians have regarded RDM as an important and valuable skill, while 27% of other information professionals have seen it as a core competency.

It is obvious that RDM requires librarians to learn new skills and abilities. For instance, Calzada Prado and Marzal (2013) emphasized the importance of knowing how to select and synthesize data and combine the data with other information sources and prior knowledge. They also enumerate the following abilities:

- To identify the context in which data is produced and reused,
- To recognize source data value, types, and formats,
- To determine when data is needed,
- To access data sources appropriate to the information needed,
- To critically assess data and their sources,
- To determine and use suitable research methods,
- To handle and analyze data,
- To present quantitative information,
- To apply results to learning, decision-making, or problem-solving,
- To plan, organize, and self-assess throughout the process.

It is not by accident that context is mentioned first in this list, because the dissociation of data from its context and the loss of context make reuse difficult or impossible (Schneider, 2013).

Sandy, Million, and Hudson-Vitale (2020, p. 204) compiled a list of knowledge, skills, and competencies, some of which are emerging ones. Many of them are more closely related to general RDM activities, while others are more library-centered. In some cases, they are tied to both. The list is as follows:

- Disciplinary knowledge,
- Knowledge of funders' policies and requirements,
- Understanding of research practices and workflows, disciplinary norms, and standards for data management,
- Advocacy and outreach,
- Data audit and assessment,
- Discovery of appropriate information sources or data repositories,
- Selection and appraisal of datasets,
- Provision of data management plan services,
- Knowledge of best practices for data structures, types, formats, vocabularies, and metadata,
- Familiarity with metadata standards,
- Data licensing and intellectual property issues,
- Data linking and data integration techniques,
- Understanding of how data is organized and structured within collections,
- Database design,
- Data preservation and archival storage,
- Data citation,
- Data sharing,
- Data visualization, and
- Data analysis.

Briney, Coates, and Goben (2020) advise to follow some other foundational practices that could help to improve RDM across all disciplines:

- Organizing files and naming them consistently,
- Applying version control (versioning),
- Creating a security plan,
- Defining roles and responsibilities,
- Providing data back-up,
- Counting with tool constraints,
- Putting the data in a repository,
- Writing the data management plan,
- Finishing a project by identifying, preparing, and collocating key research files to identify provenance and improve future retrieval.

From this list, applying version control requires a more detailed explanation. It means saving new copies of files when changes are made to them. This allows going back and retrieving specific versions of the files later, when needed. It also enables rescuing or fixing a portion of deleted material without the need of recreating it from scratch. In addition to this, keeping distinct copies of research files offers a way to document the provenance of data (Briney et al., 2020).

Fulfilling these requirements depends, among other things, on building and applying appropriate information architectures that comply with open science criteria. This indicates that the infrastructural perspective is not the only way to consider RDM.

Focusing on technical infrastructures "is the result of treating RDM as a tool to fulfil specific quality criteria stated by the open science movement" (Blask & Förster, 2020, p. 592).

According to Cox, Pinfield, and Smith (2016), RDM can be seen as a "wicked" problem by academic libraries, because it is more complex and intractable than most information problems; there are many stakeholders involved who have diverse perspectives and who do not define the associated issues in the same way. However, in some regards, the problems associated with RDM do not feel as fundamentally intractable as truly "wicked" problems. To understand this difference, some features of the concept "wicked problem" should be considered. As described by Rittel and Webber (1973), there is no definitive formulation of a wicked problem, because different approaches to a given problem are seen differently, and proposed solutions reflect these differences. Consequently, no one is in a position to say what the right answer is. There is no test to demonstrate whether a solution will work, or has worked previously. Neither is there a comprehensive list of possible solutions, and each intervention changes the problem in an irreversible way. As each wicked problem is unique, it is difficult to learn from previous problems. Moreover, a wicked problem is itself a symptom of other problems.

RDM support across large academic institutions is an area in which the library can be a driver of change and lead from the front. To accomplish this, librarians should successfully interview researchers, negotiate with other stakeholders, and demonstrate value and impact in RDM practice and expertise in relevant technology. Services provided to multiple units across the institution must be contextualized by the research life cycle (Yu et al., 2017).

In delivering RDM services, three levels of required staff participation and expertise can be discerned:

- The basic level involves activities of all public-facing staff, who are likely to be the first point of contact for researchers and are meant to foster awareness of key issues and drivers around RDM.
- Services at the advanced level are provided by select staff, who should possess tailored disciplinary knowledge and be able to provide guidance on discipline-related considerations to the users.

- Specialist RDM services rely on select staff with explicit data-related responsibilities. By having a deep understanding of the research data landscape locally, nationally, and internationally, these staff members are required to have in-depth competencies related to selected aspects of RDM (e.g., data management planning, metadata for datasets, data storage) and have explicit responsibilities around research data (Mattern, Brenner, & Lyon, 2016).

As Van Selm (2020) put it, misunderstandings should be avoided without turning away enthusiasm. We must accept that researchers do not often consider the library as the first place for getting help with data-related tasks. Neither do they have to ask for help from a librarian if they have colleagues who are able and willing to provide them with appropriate help. There are cases when the library cannot act as an expert. Nonetheless, the mission of the library dictates that librarians should become spiders in the web of data support.

Being familiar with ongoing academic discussions, researchers are expected to be experts in their own discipline. They must be familiar with new research topics and methods, and they are supposed to know which are out of fashion. They know something about information security that permits using the data now and in the future, without making the data accessible to people who have nothing to do with it, at least for as long as the data is confidential. Having knowledge of intellectual property rights is also supposed. They may fulfill these requirements, but there are, nonetheless, questions that require additional learning. For example, answering questions like "What search terms would another researcher, possibly working in a different discipline, use to find these data, and what metadata should enable funding these datasets by the right researchers?" In the ideal case, answering such questions requires librarians with 3–4 years of education, but they are trained to provide the types of services illustrated by this example. Not only digitally preserving the data but also translating many things into the language that researchers understand is the type of help provided by the library for time-pressed researchers. An example of this is explaining the differences between different Creative Commons licenses.

There is little written regarding the emotional component of research data management. Comments about how researchers feel about their data tend to appear in informal venues. Although negative emotions may be readily apparent during data management consultations, these emotions typically remain unacknowledged until trust is established (Rinehart, 2015).

Looking at characteristics of research teams offers useful particular contexts. For instance, some disciplines have open cultures with well-established standards, followed by the majority of the team's members.

Independently of the disciplinary culture, research teams have their own culture, often affected by the personality of leading researchers; thus team heterogeneity can have positive effects. Integrating researchers from different disciplines on a team may increase the likelihood that some researcher will tend to advocate for good data management practices. On a somewhat different ground, it is easier to share quantitative data that does not include information about human subjects. With datasets that need to be protected for some reason, sharing poses more challenges (Llebot & Rempel, 2021).

Planning and building RDM services

In Chapter 1, "Information, Data, Text, Document," it was indicated that academic librarians might not have a mutual understanding about their data-related mission, because of the ubiquity of the term *data* appearing in the titles of new positions and names, regarding activities that they face when forming services. This situation is aggravated by a multitude of definitions, limiting the concept of data to numbers, while others apply the word to a variety of nonnumeric resources (Swanson & Rinehart, 2016).

The results of the following studies offer a good starting point to see what deliberations can be useful when planning and carrying out RDM services.

A 2019 survey of member libraries of the Association of College & Research Libraries (ACRL) brought 186 responses, representing 5.9% of these libraries, that did not show a dramatic increase in RDM. Nonetheless, this survey by Tenopir, Kaufman, Sandusky, and Pollock (2019) revealed that a number of academic libraries offered a range of services and that the awareness of the related services had increased throughout the library community. Usually, informational (consultative) services were offered more often than technical ones. The majority of responding libraries provided reference support, followed by creating Web guides. Fifty-five of the libraries that offered technical services began these related activities around 2017. In accordance with other opinions, this survey concluded that libraries apparently do not need to offer every type of service to be active participants in RDM services. On the other hand, they should understand users' needs in order to enable tailoring services to the needs of their researchers, teaching staff members, and students.

The breakdown of informational services offered by these libraries was: Reference support (43%),

- Discussing RDM services (23%),
- Consulting on data management (21%),

- Consulting on data and metadata standards (21%),
- Training coworkers (20%),
- Being involved in data policy development (19%).

Technical services showed lower percentages:

- Technical support (26%),
- Preparing data (17%),
- Identifying data (16%),
- Creating metadata (14%),
- Deaccessioning and/or deselecting data (7%).

In their study, Gowen and Meier (2020) tracked the changes in RDM services and staffing among Association of American Universities (AAU) libraries between 2014 and 2019. They found that there were no dramatic changes in the number of libraries offering such services or employing data librarians. The only category where there was a significant increase was in the number of AAU libraries that offered a data repository. Despite losses during this period, mainly in data librarian positions, RDM support grew at these libraries.

Let us add here that, due to the apparent ambiguity of terminology for naming information professionals involved in RDM, the expression *data librarian*, or simply *librarian*, will be used, disregarding thus the perceived differences between subject librarians, subject specialists, liaison librarians, and the like.

Librarians fulfilling liaison and domain-centric roles often have opportunities to connect researchers with important information. However, they are not always well positioned to provide consultation on emerging topics. This is a main reason why they need to improve awareness and advocacy for RDM. With the changes in requirements for RDM, data sharing, and reproducible research, these librarians recognize a growing need in their work. Those librarians with appropriate domain expertise can use it for "data identification, selection, organization, preservation, and access which should be leveraged by the research data management consultants in order to provide a comprehensive service" (Wittenberg, Sackmann, & Jaffe, 2018, p. 328).

A 2018 survey of academic libraries of the Greater Western Library Alliance (GWLA) consortium in the United States garnered 35 answers out of the 39 members, covering 31 public and 8 private institutions from 20 states. The results showed that the top three topics for data workshops were general RDM, specific tools, and programming languages. The detailed answers showed that over 85% of the respondents' institutions had a repository accepting research data. In contrast to the 30% having dedicated data repositories, 70% of the institutions maintained a single repository hosting both data and other scholarly material. As for digital preservation practices, 30% of respondents only provided backup to content, 40% engaged in strategies that exceeded these backup activities, and 26% had no digital preservation strategy. Concerning metadata provision, 66% helped researchers in understanding and identifying metadata and metadata standards related to research data, but only 49% provided help for applying metadata standards to their research data (Murray et al., 2019).

The results of a nationwide online survey administered by Joo and Schmidt (2021) to academic librarians providing data services at US research universities show that many of these librarians perceive that consultation services would be more valuable to users than technical services. According to the authors, this shows that instructional skills and data management planning were perceived by many to be more important.

In the light of these data, careful planning is necessary. A series of reports devised by the Online Computer Library Center (OCLC) presents findings derived from detailed case studies of four research universities:

- The University of Edinburgh, United Kingdom;
- The University of Illinois at Urbana-Champaign, United States;
- Monash University, Australia;
- Wageningen University & Research, The Netherlands.

There are many lessons learned, but it is certain that institutions need to allocate local human resources to lead RDM initiatives, develop resources, and provide training. To accomplish this, they obviously need proper knowledge, but there is no single best model of RDM (Bryant, Lavoie, & Malpas, 2018a).

RDM services can be broken down into three types:

1. Education,
2. Expertise,
3. Curation.

At the center of educational services stands raising awareness of RDM among researchers and other stakeholders by underlining the importance and necessity of responsibly managing their data and arranging for its long-term curation.

Expertise services make use of data librarians' competencies. These services might seem to be more acceptable to researchers than educational ones, because they offer decision support and customized solutions for solving specific problems that appear during the research process. The essence of curation services is providing technical infrastructure for data management in all stages of the research cycle (Bryant, Lavoie, & Malpas, 2017a).

We must understand that "RDM is not a monolithic set of services duplicated across universities; it is a customized solution shaped by a range of internal and external factors operating on local decision-making" (Bryant, Lavoie, & Malpas, 2017b, p. 5). An RDM service bundle is not an island but is connected, to a greater or lesser degree, to the broader, external RDM ecosystem. It is also important to understand that meeting institutional needs does not necessarily require implementing the full range of services. For example, national centers can take care of local RDM needs. RDM services may not only work as a back-office component of a local research support infrastructure, but may play a more public-facing role (Bryant et al., 2017b).

Researchers' demands are crucial in shaping and sustaining RDM services, although such demands may be lacking at the initial stage of development. On the other hand, national policies may define or recommend the length of data preservation, which in turn determines institutional practice (Bryant, Lavoie, & Malpas, 2018b). In this regard, we can see that scholarly norms and data management practices are changing, but these pressures are not uniform, so the evolving practices may be different as well (Imker, 2017).

Researchers' demands appear to be particularly important as a source of intelligence in evolving RDM services, because there is uncertainty about the future of RDM, i.e., whether it can become a set of accepted scholarly practices and be part of an ecosystem of services and infrastructure. RDM used to be seen as a long-term data curation problem, which focuses on the preservation of data at the end of the research process. However, researchers' demands have changed, so it is now seen as all-encompassing management at all points of the research life cycle. (The relationship between RDM and data curation is addressed in more detail in a separate section at the end of this chapter.)

The local decisions made on research data management infrastructure, services, or personnel vary, depending on funding, research reputation, and external motivations such as scientific norms or evolving research workflows. Establishing RDM services seems therefore to be based on anticipation rather than a direct response to the demands of researchers and explicit policy mandates (Bryant et al., 2018b).

Front-end work, which includes web development and visualization, and back-end work, which includes functions like data architecture and software development, are uncommon in libraries. It has to be added here that serving the needs of academic libraries' users requires that we replace the binary (front-end and back-end) model, since embracing convergence between the two is required instead (Kowalski, 2017). Data management is an "area of expertise that blends back-of-the-house technology expertise with front-of-the-house service provision in support of the scholarly communication process" (Sandy et al., 2020, p. 199).

While libraries and other centralized units may be well positioned to act as dispatchers, there must be an understanding as to how far teaching staff members are willing to venture outside their own departments for support (Radecki & Springer, 2020). The support provided by librarians to researchers is defined not only by the areas of expertise of the scholars and the priorities of the institutions in which they work. The librarians' own skills, educational background, and training are also decisive (Sandy et al., 2020).

If we consider the scale of RDM services, we see that a number of these services are provided on a direct, one-to-one, on-demand individual consultancy basis (Bryant et al., 2017a, 2017b). Librarians also see RDM as supporting researchers' day-to-day needs and prefer providing guidance, usually to early-career researchers and in the initial phases of the research cycle (Faniel & Connaway, 2018).

Data-related roles of library research specialists include RDM, data wrangling, and data cleaning (cleansing, cleanup). Bad data quality can negatively affect decisions and have a negative impact on results. It imposes on libraries the task of being involved in data wrangling, sometimes referred to as data munging or data crunching. Its essence is first to find the dirty data, then transform and clean it. The aim of data wrangling is to prepare raw data that is restored to its original state. This makes the datasets usable for further analysis (Azeroual, 2020).

The centrality of the research data life cycle

Paying attention to changes in the global scholarly communication ecosystem and understanding the requirements set by funding agencies are important for librarians. Satisfying these requirements requires supplementing their knowledge and skills. This can be done by preparing discrete workplace tasks that individuals are expected to take on and that should make a "demonstrable difference to research practices in their institutions" (Searle, 2015, p. 6). This attention and the mission of providing integrated RDM support services also require focusing on the research life cycle.

Research data, stored together with relevant metadata, enables verification of research results and provides evidence that supports claims made by researchers. These claims are often embodied in data that can be reused for developing other research projects. These data, if cited correctly, will also increase the researchers' profile, similarly to other research outputs (Yu et al., 2017). This argument is in harmony with the thoughts about sharing and reusing data explained in Chapter 2, "Information and Data Ecologies."

Yu, Deuble, and Morgan also stated that such moves require demonstrating immediate and long-term benefits for individual researchers within the research data framework, but—if accepted—are instrumental in helping to promote RDM support services by driving cultural and behavioral change toward better RDM practices. The bottom line is that models of library services based on the research life cycle have been developed to better support research activities in their parent organizations. Using such frameworks provides an opportunity to understand where researchers are focusing their efforts in studying RDM. Life cycle models' strength is in representing and clarifying stages of research and capturing key features of project-based research. The weakness of such models is that they are closed, highly purposive, and unidirectional (Cox & Tam, 2018).

According to Perrier, Blondal, and MacDonald (2020), the related activities are creating, processing, analyzing, and preserving data; giving access to data; and reusing data. Berman (2017) focuses on the types of data collected, data file size, generation and use of metadata, and short-term and long-term data storage, data preservation, and data sharing. The first phase of the data life cycle is collection or creation, followed by processing, and reading, selecting, and reporting on the collected material. The analysis involves explaining the processes taken, asking questions, and drawing conclusions from the data (Brochu & Burns, 2019).

To understand the role of the librarian in preparing for RDM, it is important to be familiar with the research data life cycle, which is often represented in visual form. This example comes from Jisc, which is a not-for-profit organization for digital services and solutions serving higher and further education in the United Kingdom.

The activities comprising RDM, i.e., organization, storage, access, and preservation of data (Semeler, Pinto, & Rozados, 2019), are inconceivable without knowledge and skills acquired through data literacy training, because people who use research data must understand, interpret, and apply what they find, and researchers are no exception to this rule (Thomas & Urban, 2018).

We also should be aware of the synchronous nature of the research life cycle and the data management life cycle. This is because the tasks related to RDM are not entirely new just because the majority of researchers have not always been familiar with them. Earlier, they were not seen as something separated from other processes (Ojanen, Lindholm, & Siipilehto, 2020). Recognizing this fact has also changed how we see the relationship between them.

The importance of data management plans

A data management plan (DMP) is a document that provides researchers "with a mechanism for stating how they will manage data associated with at least part of a research project's data lifecycle." It is advisable to make a distinction between it and data management planning, as the former is a piece of documentation describing data management practices of a particular research project, while the latter is an active process. The first published evidence of the use of DMPs was in 1966 (Smale, Unsworth, Denyer, Magatova, & Barr, 2020, pp. 2–3). Data management plans are regarded as well-documented elements of RDM, because complying with the varied data-related requirements of research funders begins with DMPs, and assistance in preparing them is a basic level informational service that helps researchers to plan, articulate, and execute data management (Molloy & Snow, 2012; Pinfield, Cox, & Smith, 2014).

Van Loon, Akers, Hudson, and Sarkozy (2017) evaluated the content of 119 DMPs submitted to the US National Science Foundation between 2012 and 2014. The results of their content analysis indicated that most researchers understood the need for sharing data. However, many DMPs did not adequately describe how the dataset was generated, how the researchers intended to manage the data during the project, or planned to preserve and share data after the completion of the project. Obviously, the deficiencies varied across academic units. Whatever we know or think about potential positive developments in recent years, we can say that writing DMPs has remained an issue that merits the attention of researchers and often requires help from academic librarians.

DMPs formalize all technical and nontechnical details, and should be an integral part of any project that requires some kind of data management (Solodovnik & Budroni, 2015). Taking into account that the skills needed to provide assistance in producing DMPs effectively are the same as those for any informational service, such support activities require librarians to be familiar with funding requirements, relevant standards, and local data management processes (Cox & Verbaan, 2018).

Addressing several aspects of open data, the *2020 State of Open Data* report, based on a global pool of over 4500 respondents, included an expanded section on data management plans, where it is stated that the number of funders around the world mandating researchers to submit a DMP with their grant application has grown. The survey data in this report show that the number of researchers always making a DMP rose from 9% to 18%. On the other hand, the number of researchers never making a plan halved from 30% to 15%. The proportion of respondents who felt sufficiently competent to develop a practical data management plan rose to 51% (Hyndman & Goodey, 2020).

An analysis of job descriptions for library positions related to RDM, taken from several sources, identified preparing RDMs as a frequent occupation, not only because it enables researchers to gain access to tools and expertise for a better management of their research outputs, but also because it can foster conversations with institutional support service providers, such as the library (Shipman & Tang, 2019).

According to Smale et al. (2020), the potential benefits of DMPs include the following:

- Professional and institutional benefit for the researchers' productivity, visibility, integrity, and transparency;
- Economic benefits for increasing the academic and nonacademic impact of research, including the tangible and intangible "return on investment."

Nonetheless, as they also point out, there is no apparent evidence that researchers can gain professional benefit through the act of completing these documents, especially as DMPs may in fact be ineffective, because they generally do not describe research data management to any acceptable level of detail and quality. Neither is it DMPs' promulgated goal to encourage data sharing, because many researchers do not want or are not able to execute what was described in their own plans.

The inconsistent use of the concept itself makes effective data management planning difficult. This means that using outcome-focused approaches to align DMP use with specific goals by defining their explicit purpose is more than useful. This is especially true if they are written by reflecting on the well-specified concerns of an individual researcher, a specific funding agency, or an institution, instead of addressing the requirements of multiple stakeholders.

Usually, DMPs are created as free-form texts, and are often seen as an administrative exercise, instead of being regarded as an integral part of research practice. Nonetheless, as Miksa, Simms, Mietchen, and Jones (2019) put it, there are possibilities to give to DMPs not only thematic richness by indicating researchers, funders, research administrators, data librarians, and others, but also these plans can be produced in a machine-actionable form that also enables parts of the DMP to be generated and shared automatically.

The University of North Carolina at Chapel Hill, Chapel Hill, NC, University Libraries, for example, offers RDM services such as consultations on locating, acquiring, preparing, analyzing, visualizing, and preserving data. Based on data science involvement, they distinguished three main levels of involvement: data literate, data user, data-intensive (data science) researchers. Other teaching staff members (instructors) fell into the categories of nondata focused course instructors, data-science adjacent course instructors, and data-intensive/data-science course instructors. If we take the data literate category, their needs can be:

- Awareness of resources and expertise available from the library,
- Identifying potential partners/collaborators,
- Providing consultations for projects, requiring data expertise, offered by the library.

In the case of students, the following three categories deserve attention:

- Students studying in nondata-focused disciplines,
- Students focusing on a nondata-science discipline, but using data for coursework or thesis,
- Students studying a data-intensive discipline, or data science (Mani, Cawley, Henley, Triumph, & Williams, 2021).

RDM services

As was stated earlier, RDM consists of two types of services. The majority of academic libraries typically offer informational services, and technical services are also widespread. The following argument therefore is limited to some details of three informational services, related to data reference, data citation, and data retrieval.

Data reference

Just like traditional reference services, *data reference interviews* build on both the researcher and the librarian to bring their own unique expertise into the interaction. This idea is in conformity with the *Professional Competencies for Reference and User Services Librarians*, saying that a reference and user services librarian engages the user in the process and in making decisions (Huling, Dallas, Kinder, Whitlatch, & Woodard, 2017). The earlier version of this document expresses this rule more outspokenly, when stating that the librarian "acknowledges the knowledge brought by the user to the interaction" (Whitlatch, Bodner, Diefenathal Huling, & Kluegel, 2003).

Despite apparent similarities between traditional reference interviews and data reference interviews, the latter may consist of a higher number of questions than in the case of reference interactions, related to documents (texts) (Rice & Southall, 2016).

Besides assessing researcher's data needs, consultations may require the librarian to assist the researcher in finding existing datasets. Traditional skills play a role here again, as—similarly to the situation of reference librarians, who are aware of the key reference works and authoritative texts within their fields—data librarians should be familiar with major repositories and other sources of datasets (Federer, 2014). Reviewing datasets for their potential inclusion into repositories, already mentioned, is also familiar to librarians, as it shows similarities to activities rooted in collection management principles (Cox, 2018).

Data citation

Librarians also can offer help in citing datasets. Though data citation is much more complex that citing research publications, it is an unquestionable necessity (Silvello, 2018). It also has the potential to become a source of motivation for researchers to share and publish their data, if it becomes a source of reward and acknowledgment for them (Candela, Castelli, Manghi, & Tani, 2015). We know that supporting citation activities is familiar to librarians in all work environments, including data-intensive ones. Besides citing data, there is the possibility of data publication, mentioned in the chapter on ecologies, even though there is uncertainty about its precise meaning, proper forms, and extent of validating published datasets (Kratz & Strasser, 2014).

Data citation should enable giving scholarly credit and normative and legal attribution to all contributors to the data, while recognizing that a single style or mechanism of attribution may not be applicable to all data. It also has to be directed toward facilitating access to the data itself and to associated metadata, documentation, code, and other materials. Identifying access and verifying specific data that support a claim must be attainable. Providing unique identifiers and metadata can describe the data in a persistent way, even beyond the lifespan of the data that is described (Data Citation Synthesis Group, 2014).

According to Borgman (2016, p. 4), these principles fulfill the same role as bibliographic citation, i.e., "documenting evidence, accommodating variant practices among communities, identifying cited items unambiguously, and improving access to the cited objects." Obviously, data citations should be in digital form, available online, and must support machine action. Besides this, to make citations interpretable, data must be combined with contextual information. Indicating the provenance of data is also a necessity. While publications are presumed to be interpretable as independent units, data may be interpretable only in combination with contextual information.

Data retrieval

Beyond assisting researchers in assigning metadata and citations to datasets, helping them in *data retrieval* is also a potential task. However, data retrieval is a less obvious task than retrieving research papers, which are available in a predominantly textual form. For instance, the average file size of retrieved datasets is several times larger than that of retrieved research publication files; and these in turn vary from one discipline to another. In addition, the retrieved datasets may be of different file types or formats and often must be downloaded before they can be read or used. Research datasets are usually accompanied by separate documentation files and a single dataset item record may constitute several composite files (Bugaje & Chowdhury, 2018).

As Ma (2013) explains, there are differences between retrieving documents and big data. Information retrieval systems, including search engines, depend on the manipulation of encoded data. In this context, what is called *data* usually refers to a text-based or graphic document. Nonetheless, retrieval is done without making sense of these documents. Instead, data is transformed into information by involving some kind of representation, e.g., bibliographic records, classification systems, or taxonomical structures. However, we can presume that datasets provide only a vague picture of their content and meaning; therefore it is impossible to create useful representations. Simulations of the objective reality are created instead. Instead of having their own interpretation and understanding, users get presumably true and real packages of facts.

There are further differences as well. As Bugaje and Chowdhury (2018) state, the fundamental characteristics of research data and the user interactions that it requires are different from those of publishing papers and books; thus adapting standard text retrieval to data is impractical. The reason for this is that—although both texts and datasets can be tagged with metadata—this task is more complex in the case of research data. Similarly, in contrast to indexing research papers, the indexing of research datasets lacks standardization or control.

The average file size of datasets is usually larger than that of research publication files. In addition to this, the retrieved datasets may contain different file types and/or file formats and may require downloading in order to be readable or used. In contrast to research publications that comprise only the publication itself, datasets are usually accompanied by separate documentation files. To this comes the fact that single dataset items may contain several composite files, comprising fragments or versions of the dataset.

Furthermore, we also have to face the problem that the currently available data retrieval services are unsustainable, as they involve an unnecessarily high consumption of valuable resources. This includes increased network and researcher time due to the already named fact that research datasets must be downloaded to enable their contents to be viewed. These unnecessary downloads involve not only high amounts of energy consumption with environmental and economic implications, but also comprise additional server costs. As these downloads may prove empty, they are a hindrance in judging the reliability of data.

RDM and data curation

In the support chain of researchers, data literacy instruction is often the first step, followed by RDM. The third activity is data curation (DC), which may be compared to curating a museum collection, where pieces are exhibited rather than kept for internal storage, whereas data needs to be preserved by frequent validation checks and backups (Thomas & Urban, 2018).

Data curation must be a long-term engagement of which technology is only one of the components, because—due to the heterogeneity of data-related roles, stemming from complex social and behavioral systems—the involvement of humans is a necessity (Poole, 2015).

Curation involves planning, acquiring, preparing, analyzing, preserving, and discovering the data (Pouchard, 2016). The preservation of digital information is a complex and costly process that cannot be circumvented (Burgi, Blumer, & Makhlouf-Shabou, 2017).

Data curation highlights the need for abandoning opinions, which portray data as something inferior to information (Cox, 2018). These and similar views may sound familiar from the argument of Chapter 1, if not elsewhere. DC is often seen as a part of RDM that reflects the views on its nature. It is also a set of general activities, such as organizing, collecting, describing, and storing data for communication among scholars, which is not specifically attached to the library, but potentially performed by it (Thomas & Urban, 2018).

Data curation can also be defined as the "encompassing work and actions taken by curators of a data repository, in order to provide meaningful and enduring access to data." In their turn, data curation services provide a technical infrastructure that supports data management by providing persistent storage and assignment of unique identifiers and metadata (Johnston et al., 2018, p. 5).

At the center of data curation, therefore, stands preservation for reusability, if we share the opinion that the value of data is exclusively in its extrinsic properties, i.e., the fitness for use (Sposito, 2017). In this sense, research data differs from digitized cultural heritage objects, which hold intrinsic value that justifies preservation regardless of ever being utilized or not.

Data curation is not only a task for libraries and other nonprofit organizations. There is also commercial interest in hosting data and providing access to it. However, this may be limited to hosting data without readiness for curating data. As mentioned in Chapter 2 on information and data ecologies, data journals and journals that publish research data attached to papers, or even citable and cross-linked data papers, show that publishers are interested in providing access to data as value-added services, but this does not mean that they would be willing to host data, except as a for-profit venture. Data repositories, typically run by nonprofit consortia, care for their long-term ability to curate data resources. Universities are interested both in managing their resources, because the records of the scholarly output of their researchers and students can be used for promoting their reputation and allow evaluating people's and departments' achievements and prestige. This can be carried out by maintaining data repositories, among other methods. Academic libraries see a potential role in curating data produced by researchers in their universities or other organizations, but may not wish to compete with repositories. On the other hand, they are willing to apply their expertise in information organization and curation (Borgman, 2016).

When promoting DC, academic libraries should accentuate that they will be responsible stewards of data, without taking ownership of this data. This means that the owners of any material will be able to take their content out of library systems at any point and that this content will be accessible (to the best of libraries' abilities) in perpetuity (Lucky & Harkema, 2018).

The skills and expertise required to prepare, arrange, describe, and optimize data for reuse cannot be fully automated. Curation staff therefore builds the "human layer" by bringing in disciplinary knowledge and software expertise. Curating data should not be siloed at a single institution. Moreover, data curation experts should ensure that the incoming datasets comply with the FAIR principles (described in Chapter 2) (Johnston et al., 2018).

Librarians have acquired knowledge and skills in managing libraries as organized collections within a certain place and space. These skills may prove useful for data curation, as well. We can therefore regard them as qualified professionals who can be entrusted by making use of their competencies (Burgi et al., 2017). However, curating digital data is not their exclusive domain.

There are also curators working in a variety of institutions and caring for a host of artifacts. Somewhat differently, archivists care for the permanency of records and their historical value. Data curation's main goals overlap with those of digital curation, because both of them aim at preventing data loss and endeavor to add value to trusted data assets for current and future use (Poole, 2016). This can be seen also in their definitions that emphasize the processes related to taking care of resources, including organization and description (Kouper, 2016). As Robinson (2016) put this, if we regard data as *any information in digital form* (sic!), then data curation is identical with digital curation, i.e., they both consist of the management and preservation of digital material.

According to a classic definition of Lee and Tibbo (2007), digital curation "involves selection and appraisal by creators and archivists; evolving provision of intellectual access; redundant storage; data transformations; and, for some materials, a commitment to long-term preservation. Digital curation is stewardship that provides for the reproducibility and re-use of authentic digital data and other digital assets. Development of trustworthy and durable digital repositories; principles of sound metadata creation and capture;

use of open standards for file formats and data encoding; and the promotion of information management literacy are all essential to the longevity of digital resources and the success of curation efforts."

Johnston et al. (2017) examined six US academic library institutions' data repository and curation services and found that each curation workflow is based on a self-submission model that allows researchers to deposit their data at will, and all repositories (except for one) automatically accepted these data. Digital object identifiers (DOIs) are used as persistent identifiers, but these data curation services augment the metadata and documentation added to the record.

The US Data Curation Network examined 114 ARL institutions and reported that about 44% of them had a dedicated data repository, but information about data curation support is rarely available on these websites (Johnston & Coburn, 2020).

In an online survey of professionals who identify themselves as digital curators, Kouper (2016) found preservation as the most frequent theme. While data curation, similarly to data literacy and RDM, has to focus on data quality (Ridsdale et al., 2015), it goes beyond RDM, because it additionally comprises preservation and enables reuse (Partlo, Symons, & Carlson, 2015).

In order to be successful, data curation requires management of digital data over its entire life cycle, including the use of varied services, tools, and infrastructure (Schmidt & Shearer, 2016) to ensure the maintenance of authenticity, reliability, integrity, and usability of digital material, which in turn ensures maximization of the investment in their creation (Higgins, 2008, p. 135). By subsuming digital preservation and supplying different kinds of documentation to provide context, as well as including metadata, data curation aims to make selected data accessible, usable, and useful throughout its life cycle (Giarlo, 2013).

RDM and research support services

Research support is a broad concept, broader than RDM, often reflecting the broader perspective of open access and open science. Support activities are on their way to being transformed into eResearch roles, impersonating in many regards the role model of the data librarian (Brown, Wolski, & Richardson, 2015). A few case studies illustrate the variety of research support services offered by academic libraries. Let us add that these services are not limited to research data. Nonetheless, the development of new services at the Griffith University Library (Brisbane, Australia) has been described by Weaver and Richardson (2020). The researcher support services at this university are designed to support data-driven, data-intensive research. These services assist researchers through the entire research life cycle. Their team addresses the research data life cycle and provides clarity about licensing, data sharing and reuse, and other data-related issues. Services also include data storytelling and tools for visualization. Staff members are being trained in how to conduct data interviews, about the optimal colocation of data and tools, and methods for remote collaboration over data. The range of services listed in their service catalogue included data management, scholarly publishing (including academic writing) and research impact, open scholarship, and data wrangling. Data wrangling was a new, in essence, RDM activity among these areas, which has been built up completely from scratch. The diversity of tasks is also often dominant. In a report summarizing the discussions and major themes that emerged from a workshop on Developing the Librarian Data Science and Open Science, Federer, Clarke, and Zaringhalam (2020) also provide a similarly colorful selection of tasks, with bibliometrics and research impact assessments performed by a librarian, including database searching, report writing, data analysis, data visualization, and data wrangling.

At Kaunas University of Technology (Lithuania), one of the training initiatives led by the library was developing an elective doctoral RDM course in collaboration with the research office, doctoral school, and the IT department. This course addressed related guidelines, standards, and tools, legal and ethical issues, data licensing, data citations, data sharing, repositories, and archives. Preparing DMPs was an obligatory element of the curriculum. Open science initiatives receive special emphasis as the University became the National Node of the Research Data Alliance (RDA) Europe. In 2019, they started a 1-year project that aimed to raise awareness about and support for RDM in Lithuania (Cesevičiūtė & Tautkevičienė, 2020).

The library of Te Herenga Waka—Victoria University of Wellington, New Zealand offered a Research Data Management subject guide. The research support librarians were supported by the team of subject librarians. Their self-development included using the 23 (research data) Things, which is a self-directed, learning program (https://www.ands.org.au/working-with-data/skills/23-research-data-things) (Wilson, Kirkpatrick, Miller, & Harvey, 2020).

Conclusion

The need for research data management is not only a new but also a complex issue. The readiness for providing the corresponding services has been varied, but their recognition seems to be growing. Data is produced by the researchers themselves, moreover they also curate their datasets. Nonetheless, academic libraries can and should play a central role in managing and curating data, and must continually develop their services, similarly to the way described in this chapter.

References

ACRL. (2012). 2012 top ten trends in academic libraries: A review of the trends and issues affecting academic libraries in higher education. *College & Research Libraries News, 73*(6), 311–320. https://doi.org/10.5860/crln.73.6.8773.

ACRL. (2014). Top ten trends in academic libraries. A review of the trends and issues affecting academic libraries in higher education. *College & Research Libraries News, 75*(6), 294–302. https://doi.org/10.5860/crln.75.6.9137.

ACRL. (2016). 2016 top trends in academic libraries. A review of the trends and issues affecting academic libraries in higher education. *College & Research Libraries News, 77*(6), 274–281. https://doi.org/10.5860/crln.77.6.9505.

ALLEA (2017) European Science Foundation and All European Academies. (2017). *The European code of conduct for research integrity*. European Science Foundation. Retrieved from https://www.allea.org/wp-content/uploads/2017/05/ALLEA-European-Code-of-Conduct-for-Research-Integrity-2017.pdf.

Azeroual, O. (2020). Data wrangling in database systems: Purging of dirty data. *Data, 5*(2), 50. https://doi.org/10.3390/data5020050.

Benedetti, A., Boehme, G., Caswell, T. R., Denlinger, K., Li, Y., McAllister, A. D., ... Wesolek, A. J. (2020). 2020 top trends in academic libraries. *College & Research Libraries News, 81*(6), 270–278. https://doi.org/10.5860/crln.81.6.270.

Berman, E. A. (2017). An exploratory sequential mixed methods approach to understanding researchers' data management practices at UVM: Integrated findings to develop research data services. *Journal of eScience Librarianship, 6*(1), e1104. https://doi.org/10.7191/jeslib.2017.1104.

Blask, K., & Förster, A. (2020). Designing an information architecture for data management technologies: Introducing the DIAMANT model. *Journal of Librarianship and Information Science, 52*(2), 592–600. https://doi.org/10.1177/0961000619841419.

Borgman, C. L. (2016). Data citation as a bibliometric oxymoron. In C. R. Sugimoto (Ed.), *Theories of informetrics and scholarly communication* (pp. 93–115). Berlin, Boston: Walter de Gruyter GmbH & Co KG.

Briney, K. A., Coates, H. L., & Goben, A. (2020). Foundational practices of research data management. *Research Ideas and Outcomes, 6*, e56508. https://doi.org/10.3897/rio.6.e56508.

Brochu, L., & Burns, J. (2019). Librarians and research data management—A literature review: Commentary from a senior professional and a new professional librarian. *New Review of Academic Librarianship, 25*(1), 49–58. https://doi.org/10.1080/13614533.2018.1501715.

Brown, R. A., Wolski, M., & Richardson, J. (2015). Developing new skills for research support librarians. *The Australian Library Journal, 64*(3), 224–234. https://doi.org/10.1080/00049670.2015.1041215.

Bryant, R., Lavoie, B., & Malpas, C. (2017a). Scoping the university RDM service bundle. In *The realities of research data management, part 2*. Dublin, OH: OCLC Research. https://doi.org/10.25333/C3Z039.

Bryant, R., Lavoie, B., & Malpas, C. (2017b). *A tour of the research data management (RDM) service space. the realities of research data management, part 1*. Dublin, OH: OCLC Research. https://doi.org/10.25333/C3PG8J.

Bryant, R., Lavoie, B., & Malpas, C. (2018a). Sourcing and scaling university RDM services. In *The realities of research data management, part 4*. Dublin, OH: OCLC Research.

Bryant, R., Lavoie, B., & Malpas, C. (2018b). Incentives for building university RDM services. In *The realities of research data management, part 3*. Dublin, OH: OCLC Research. https://doi.org/10.25333/C3S62F.

Bugaje, M., & Chowdhury, G. (2018). Identifying design requirements of a user-centered research data management system. In M. Dobreva, A. Hinze, & M. Žumer (Eds.), *Maturity and innovation in digital libraries* (pp. 335–347). Cham: Springer. https://doi.org/10.1007/978-3-030-04257-8_35.

Burgi, P. Y., Blumer, E., & Makhlouf-Shabou, B. (2017). Research data management in Switzerland: National efforts to guarantee the sustainability of research outputs. *IFLA Journal*, *43*(1), 5–21. https://doi.org/10.1177/2F0340035216678238.

Calzada Prado, J. C., & Marzal, M.Á. (2013). Incorporating data literacy into information literacy programs: Core competencies and contents. *Libri*, *63*(2), 123–134. https://doi.org/10.1515/libri-2013-0010.

Candela, L., Castelli, D., Manghi, P., & Tani, A. (2015). Data journals: A survey. *Journal of the Association for Information Science and Technology*, *66*(9), 1747–1762. https://doi.org/10.1002/asi.23358.

Cesevičiūtė, I., & Tautkevičienė, G. (2020). Research data management support at Kaunas University of Technology. In *Cases on research support services in academic libraries* (pp. 52–71). Heshey, PA: IGI Global. https://doi.org/10.4018/978-1-7998-4546-1.ch003.

Chawinga, W. D., & Zinn, S. (2020). Research data management in universities: A comparative study from the perspectives of librarians and management. *The International Information and Library Review*. https://doi.org/10.1080/10572317.2020.1793448.

Chiware, E. R., & Becker, D. A. (2018). Research data management services in Southern Africa: A readiness survey of academic and research libraries. *African Journal of Library, Archives and Information Science*, *28*(1), 1–16. https://doi.org/10.7553/81-2-1563.

Christensen-Dalsgaard, B., van den Berg, M., Grim, R., Horstmann, W., Jansen, D., Pollard, T., & Roos, A. (2012). *Ten recommendations for libraries to get started with research data management*. LIBER. http://www.libereurope.eu/news/ten-recommendations-for-libraries-to-get-started-with-research-data-management.

Cox, A. (2018). Academic librarianship as a data profession. In *Information Today Europe/ILI365*. https://www.infotoday.eu/Articles/Editorial/Featured-Articles/Academic-librarianship-as-a-data-profession-125376.aspx.

Cox, A. M., Kennan, M. A., Lyon, L., & Pinfield, S. (2017). Developments in research data management in academic libraries: Towards an understanding of research data service maturity. *Journal of the Association for Information Science and Technology*, *68*(9), 2182–2200. https://doi.org/10.1002/asi.23781.

Cox, A. M., Pinfield, S., & Smith, J. (2016). Moving a brick building: UK libraries coping with research data management as a 'wicked' problem. *Journal of Librarianship and Information Science*, *48*(1), 3–17. https://doi.org/10.1177/0961000614533717.

Cox, A. M., & Tam, W. W. T. (2018). A critical analysis of lifecycle models of the research process and research data management. *Aslib Journal of Information Management*, *70*(2), 142–157. https://doi.org/10.1108/AJIM-11-2017-0251.

Cox, A. M., & Verbaan, E. (2018). *Exploring research data management*. London: Facet.

Data Citation Synthesis Group. (2014). *Joint declaration of data citation principles*. San Diego, CA: FORCE11. https://doi.org/10.25490/a97f-egyk. Retrieved from.

Dempsey, L. (2017). Library collections in the life of the user: Two directions. *LIBER Quarterly*, *26*(4), 338–359. https://doi.org/10.18352/lq.10170.

Ducas, A., Michaud-Oystryk, N., & Speare, M. (2020). Reinventing ourselves: New and emerging roles of academic librarians in Canadian research-intensive universities. *College & Research Libraries*, *81*(1), 43–65. https://doi.org/10.5860/crl.81.1.43.

Faniel, I. M., & Connaway, L. S. (2018). Librarians' perspectives on the factors influencing research data management programs. *College and Research Libraries*, *79*(1), 100–119. https://doi.org/10.5860/crl.79.1.100.

Federer, L. (2014). *Emerging trends in librarianship: Exploring new roles for librarians: The research informationist*. San Rafael, CA: Morgan & Claypool Publishers.

Federer, L., Clarke, S. C., & Zaringhalam, M. (2020). *Developing the librarian workforce for data science and open science*. Bethesda, MD: National Library of Medicine. https://osf.io/uycax/.

Giarlo, M. J. (2013). Academic libraries as data quality hubs. *Journal of Librarianship and Scholarly Communication*, *1*(3), eP1059. https://doi.org/10.7710/2162-3309.1059.

Gowen, E., & Meier, J. J. (2020). Research data management services and strategic planning in libraries today: A longitudinal study. *Journal of Librarianship and Scholarly Communication, 8*(General Issue), eP2336. https://doi.org/10.7710/2162-3309.2336.

Higgins, S. (2008). The DCC curation lifecycle model. *International Journal of Digital Curation, 3*(1), 134–140.

Hinrichs-Krapels, S., & Grant, J. (2016). Exploring the effectiveness, efficiency and equity (3e's) of research and research impact assessment. *Palgrave Communications, 2*(1), 1–9. https://doi.org/10.1057/palcomms.2016.90.

Hsiao, T. M., & Chen, K. H. (2020). The dynamics of research subfields for library and information science: An investigation based on word bibliographic coupling. *Scientometrics, 125*(1), 717–737. https://doi.org/10.1007/s11192-020-03645-9.

Huling, N., Dallas, L. J., Kinder, R., Whitlatch, J. B., & Woodard, B. (2017). *Professional competencies for reference and user services librarians*. Chicago, IL: American Library Association. Retrieved from http://www.ala.org/rusa/resources/guidelines/professional.

Hyndman, A., & Goodey, G. (2020). It all starts with a good plan. In *The state of open data* (pp. 17–21). Cham: Springer Nature. https://doi.org/10.6084/m9.figshare.13227875.

Imker, H. J. (2017). Overlooked and overrated data sharing: Why some scientists are confused and/or dismissive. In L. R. Johnston (Ed.), *Practical strategies for your digital repository: Vol. 1. Curating research data* (pp. 127–150). Chicago, IL: Association of College and Research Libraries. http://hdl.handle.net/2142/95024.

Jahnke, L., Asher, A., & Keralis, S. D. (2012). *The problem of data*. Washington, DC: Council on Library and Information Resources.

Johnston, L. R., Carlson, J., Hswe, P., Hudson-Vitale, C., Imker, H., Kozlowski, W., … Stewart, C. (2017). Data curation network: How do we compare? A snapshot of six academic library institutions' data repository and curation services. *Journal of eScience Librarianship, 6*(1), e1102. https://doi.org/10.7191/jeslib.2017.1102.

Johnston, L. R., Carlson, J., Hudson-Vitale, C., Imker, H., Kozlowski, W., Olendorf, R., & Hull, E. (2018). Data curation network: A cross-institutional staffing model for curating research data. *International Journal of Digital Curation, 13*(1), 125–140. 125 https://doi.org/10.2218/ijdc.v13i1.616.

Johnston, L. R., & Coburn, L. (2020). *Data sharing readiness in academic institutions*. Data Curation Network. University of Minnesota Digital Conservancy. http://hdl.handle.net/11299/211358.

Joo, S., & Schmidt, G. M. (2021). Research data services from the perspective of academic librarians. *Digital Library Perspectives*. https://doi.org/10.1108/DLP-10-2020-0106.

Koltay, T. (2016). Are you ready? Tasks and roles for academic libraries in supporting Research 2.0. *New Library World, 117*(1/2), 94–104. https://doi.org/10.1108/NLW-09-2015-0062.

Koltay, T. (2019a). Identifying new roles for academic libraries in supporting data-intensive research. *Bibliosphere, 4*, 97–102. https://doi.org/10.20913/1815-3186-2019-4-97-102.

Koltay, T. (2019b). Data curation in academic libraries as part of the digital revolution. *Zagadnienia Informacji Naukowej–Studia Informacyjne, 57*(1A (113A)), 28–36.

Kouper, I. (2016). Professional participation in digital curation. *Library & Information Science Research, 38*(3), 212–223. https://doi.org/10.1016/j.lisr.2016.08.009.

Kowalski, M. (2017). Breaking down silo walls: Successful collaboration across library departments. *Library Leadership & Management, 2*(31), 1–15. https://doi.org/10.5860/llm.v31i2.7202.

Kratz, J., & Strasser, C. (2014). Data publication consensus and controversies. *F1000Research*, 3. https://doi.org/10.12688/f1000research.3979.3.

Lee, C. A., & Tibbo, H. R. (2007). Digital curation and trusted repositories: Steps toward success. *Journal of Digital Information, 8*(2), 34. Retrieved from https://journals.tdl.org/jodi/index.php/jodi/issue/view/34.

LERU. (2013). *LERU roadmap for research data*. Leuven: League of European Research Universities Research Data Working Group. Retrieved from http://www.leru.org/index.php/public/publications/year/2013/.

Llebot, C., & Rempel, H. G. (2021). Why won't they just adopt good research data management practices? An exploration of research teams and librarians' role in facilitating RDM adoption. *Journal of Librarianship and Scholarly Communication, 9*(General Issue), eP2321. https://doi.org/10.7710/2162-3309.2321.

Lucky, S., & Harkema, C. (2018). Back to basics: Supporting digital humanities and community collaboration using the core strength of the academic library. *Digital Library Perspectives, 34*(3), 188–199. https://doi.org/10.1108/DLP-03-2018-0009.

Ma, L. (2013). Is information still relevant? *Information Research, 18*(3). paper C33. Retrieved from http://InformationR.net/ir/18-3/colis/paperC33.html.

Mani, N. S., Cawley, M., Henley, A., Triumph, T., & Williams, J. M. (2021). Creating a data science framework: A model for academic research libraries. *Journal of Library Administration, 61*(3), 281–300. https://doi.org/10.1080/01930826.2021.1883366.

Martell, C. (2009). sAccess: The social dimension of a new paradigm for academic librarianship. *Journal of Academic Librarianship, 35*(3), 205–206. doi:10.1016%2Fj.acalib.2009.03.001.

Mattern, E., Brenner, A., & Lyon, L. (2016). Learning by teaching about RDM: An active learning model for internal library education. *International Journal of Digital Curation, 11*(2), 27–38. https://doi.org/10.2218/ijdc.v11i2.414.

Miksa, T., Simms, S., Mietchen, D., & Jones, S. (2019). Ten principles for machine-actionable data management plans. *PLoS Computational Biology, 15*(3), e1006750. https://doi.org/10.1371/journal.pcbi.1006750.

Molloy, L., & Snow, K. (2012). The data management skills support initiative: Synthesising postgraduate training in research data management. *International Journal of Digital Curation, 7*(2), 101–109. https://doi.org/10.2218/ijdc.v7i2.233.

Murray, M., O'Donnell, M., Laufersweiler, M. J., Novak, J., Rozum, B., & Thompson, S. (2019). A survey of the state of research data services in 35 U.S. academic libraries, or "wow, what a sweeping question". *Research Ideas and Outcomes, 5*, e48809. https://doi.org/10.3897/rio.5.e48809.

NMC. (2014). *NMC horizon report: 2014 library edition*. Austin, TX: The New Media Consortium.

Ojanen, M., Lindholm, T., & Siipilehto, L. (2020). *What is research data management (RDM)?*. Retrieved from https://blogs.helsinki.fi/thinkopen/know-your-data-rdm-series-1/.

Partlo, K., Symons, D., & Carlson, J. D. (2015). Revolutionary or evolutionary? Making research data management manageable. In B. L. Eden (Ed.), *Creating research infrastructures in the 21st century academic library: Conceiving, funding, and building new facilities and staff* (pp. 175–201). Lanham, MD: Rowman and Littlefield.

Perrier, L., Blondal, E., & MacDonald, H. (2018). Exploring the experiences of academic libraries with research data management: A meta-ethnographic analysis of qualitative studies. *Library & Information Science Research, 40*(3–4), 173–183. https://doi.org/10.1016/j.lisr.2018.08.002.

Perrier, L., Blondal, E., & MacDonald, H. (2020). The views, perspectives, and experiences of academic researchers with data sharing and reuse: A meta-synthesis. *PLoS ONE, 15*(2), e0229182. https://doi.org/10.1371/journal.pone.0229182.

Pinfield, S., Cox, A. M., & Smith, J. (2014). Research data management and libraries: Relationships, activities, drivers and influences. *PLoS One, 9*(12), e114734. https://doi.org/10.1371/journal.pone.0114734.

Poole, A. H. (2015). How has your science data grown? Digital curation and the human factor: A critical literature review. *Archival Science, 15*(2), 101–139. https://doi.org/10.1007/s10502-014-9236-y.

Poole, A. H. (2016). The conceptual landscape of digital curation. *Journal of Documentation, 72*(5), 961–986. https://doi.org/10.1108/JD-10-2015-0123.

Pouchard, L. (2016). Revisiting the data lifecycle with big data curation. *International Journal of Digital Curation, 10*(2), 176–192. https://doi.org/10.2218/ijdc.v10i2.342.

Radecki, J., & Springer, R. (2020). *Research data services in US higher education: A web-based inventory*. New York, NY: Ithaka S + R.

Rice, R., & Southall, J. (2016). *The data librarian's handbook*. London: Facet Publishing.

Ridsdale, C., Rothwell, J., Smit, M., Ali-Hassan, H., Bliemel, M., & Irvine, D. (2015). *Strategies and best practices for data literacy education. Knowledge synthesis report*. Halifax, NS: Dalhousie University.

Rinehart, A. K. (2015). Getting emotional about data. The soft side of data management services. *College & Research Libraries News*, *76*(8), 437–440. https://doi.org/10.5860/crln.76.8.9364.

Rittel, H. W., & Webber, M. M. (1973). Dilemmas in a general theory of planning. *Policy Sciences*, *4*(2), 155–169.

Robinson, L. (2016). Between the deluge and the dark age: Perspectives on data curation. *Alexandria*, *26*(2), 73–76.

Robinson, L., & Bawden, D. (2017). 'The story of data': A socio-technical approach to education for the data librarian role in the CityLIS library school at City, University of London. *Library Management*, *38*(6/7), 312–322. https://doi.org/10.1108/LM-01-2017-0009.

Sandy, H. M., Million, A. J., & Hudson-Vitale, C. (2020). Innovating support for research: The coalescence of scholarly communication? *College & Research Libraries*, *81*(2), 193–213. https://doi.org/10.5860/crl.81.2.193.

Saunders, L. (2020). Core knowledge and specialized skills in academic libraries. *College & Research Libraries*, *81*(2), 288–311. https://doi.org/10.5860/crl.81.2.288.

Sayre, F., & Riegelman, A. (2019). Replicable services for reproducible research: A model for academic libraries. *College & Research Libraries*, *80*(2), 260–272. https://doi.org/10.5860/crl.80.2.260.

Schmidt, K., & Shearer, B. S. K. (2016). *Librarians competencies profile for research data management*. Joint Task Force on Librarians' Competencies in Support of EResearch and Scholarly Communication. https://www.coar-repositories.org/files/Competencies-for-ScholComm-and-OA_June-2016.pdf.

Schneider, R. (2013). Research data literacy. In S. Kurbanoglu, et al. (Eds.), *Worldwide commonalities and challenges in information literacy research and practice* (pp. 134–140). Springer International.

Searle, S. (2015). Using scenarios in introductory research data management workshops for library staff. *D-Lib Magazine*, *21*(11). https://doi.org/10.1045/november2015-searle.

Semeler, A. R., & Pinto, A. L. (2020). Data librarianship as a field. *Transinformação*, *32*, e200034. https://doi.org/10.1590/2318-0889202032e200034.

Semeler, A. R., Pinto, A. L., & Rozados, H. B. F. (2019). Data science in data librarianship: Core competencies of a data librarian. *Journal of Librarianship and Information Science*, *51*(3), 771–780. https://doi.org/10.1177/0961000617742465.

Sesartić, A., Fischlin, A., & Töwe, M. (2016). Towards narrowing the curation gap—Theoretical considerations and lessons learned from decades of practice. *ISPRS International Journal of Geo-Information*, *5*(6), 91. https://doi.org/10.3390/ijgi5060091.

Shipman, J. P., & Tang, R. (2019). The collaborative creation of a Research Data Management Librarian Academy (RDMLA). *Information Services & Use*, *39*(3), 243–247. https://doi.org/10.3233/ISU-190050.

Silvello, G. (2018). Theory and practice of data citation. *Journal of the Association for Information Science and Technology*, *69*(1), 6–20. https://doi.org/10.1002/asi.23917.

Smale, N. A., Unsworth, K., Denyer, G., Magatova, E., & Barr, D. (2020). A review of the history, advocacy and efficacy of data management plans. *International Journal of Digital Curation*, *15*(1). https://doi.org/10.2218/ijdc.v15i1.525.

Solodovnik, I., & Budroni, P. (2015). Preserving digital heritage. At the crossroads of Trust and Linked Open Data. *IFLA Journal*, *41*(3), 251–264. https://doi.org/10.1177/0340035215600453.

Sposito, F. A. (2017). *What do data curators care about? Data quality, user trust, and the data reuse plan*. Paper presented at IFLA WLIC 2017. Retrieved from http://library.ifla.org/1797/.

Swanson, J., & Rinehart, A. K. (2016). Data in context: Using case studies to generate a common understanding of data in academic libraries. *Journal of Academic Librarianship, 42*(1), 97–101. https://doi.org/10.1016/j.acalib.2015.11.005.

Tang, R., & Hu, Z. (2019). Providing Research Data Management (RDM) services in libraries: Preparedness, roles, challenges, and training for RDM practice. *Data and Information Management, 3*(2), 84–101. https://doi.org/10.2478/dim-2019-0009.

Tenopir, C., Kaufman, J., Sandusky, R. J., & Pollock, D. (2019). *Research data Services in academic libraries: Where are we today?*. Chicago, IL: ACRL Choice. Retrieved from https://www.choice360.org/research/research-data-services-in-academic-libraries-where-are-we-today/.

Thomas, C. V., & Urban, R. J. (2018). What do data librarians think of the MLIS? Professionals' perceptions of knowledge transfer, trends, and challenges. *College & Research Libraries, 79*(3), 401–423. https://doi.org/10.5860/crl.79.3.401.

TU/e, n.d. Technische Universiteit Eindhoven. *What is research data management?* Retrieved from https://www.tue.nl/en/our-university/library/education-research-support/scientific-publishing/data-coach/general-terms-and-background/what-is-research-data-management/.

Van Loon, J. E., Akers, K. G., Hudson, C., & Sarkozy, A. (2017). Quality evaluation of data management plans at a research university. *IFLA Journal, 43*(1), 98–104. https://doi.org/10.1177/0340035216682041.

Van Selm, M. (2020). *If a researcher would meet a librarian... The state of open data* (pp. 4–16). London, Belin, New York, NY: Springer Nature. https://doi.org/10.6084/m9.figshare.13227875.

Vilar, P., & Zabukovec, V. (2019). Research data management and research data literacy in Slovenian science. *Journal of Documentation, 75*(1), 24–43.

WCRIF. (2019). *Hong Kong Principles*. World Conference on Research Integrity Foundation. Retrieved from https://wcrif.org/guidance/hong-kong-principles.

Weaver, B., & Richardson, J. (2020). Reinventing library research support services at Griffith University. In *Cases on research support services in academic libraries* (pp. 267–289). Heshey, PA: IGI Global. https://doi.org/10.4018/978-1-7998-4546-1.ch012.

Whitlatch, J. B., Bodner, N. E., Diefenathal Huling, N., & Kluegel, K. M. (2003). Professional competencies for reference and user services librarians. *Reference and User Services Quarterly, 42*(4), 290–295.

Willaert, T., Cottyn, J., Kenens, U., Vandendriessche, T., Verbeke, D., & Wyns, R. (2019). Research data management and the evolutions of scholarship: Policy, infrastructure and data literacy at KU Leuven. *LIBER Quarterly, 29*(1). https://doi.org/10.18352/lq.10272.

Wilson, T., Kirkpatrick, A., Miller, K., & Harvey, M. R. (2020). Beyond the crossroads: Transforming research support services at Te Herenga Waka Library. In *Cases on research support services in academic libraries* (pp. 290–311). Heshey, PA: IGI Global. https://doi.org/10.4018/978-1-7998-4546-1.ch013.

Wittenberg, J., Sackmann, A., & Jaffe, R. (2018). Situating expertise in practice: Domain-based data management training for liaison librarians. *The Journal of Academic Librarianship, 44*(3), 323–329. https://doi.org/10.1016/j.acalib.2018.04.004.

Yu, F., Deuble, R., & Morgan, H. (2017). Designing research data management services based on the research lifecycle—A consultative leadership approach. *Journal of the Australian Library and Information Association, 66*(3), 287–298. https://doi.org/10.1080/24750158.2017.1364835.

Zhang, L., & Eichmann-Kalwara, N. (2019). Mapping the scholarly literature found in Scopus on "research data management": A bibliometric and data visualization approach. *Journal of Librarianship and Scholarly Communication, 7*(1), eP2226. https://doi.org/10.7710/2162-3309.2226.

Further reading

FAIR Data Maturity Model Working Group. (2020). *FAIR data maturity model: Specification and guidelines*. RDA. https://doi.org/10.15497/rda00050.

Ngoc-Yen, T., & Chan, E. K. (2020). *Opportunities for academic libraries*. ACRL/Choice Publisher. Retrieved from https://www.choice360.org/research/supporting-scholarly-research-current-and-new-opportunities-for-academic-libraries/RDA.

Tenopir, C., Kaufman, J., Sandusky, R. J., & Pollock, D. E. (2020). *Research data services in academic libraries: Where are we today?*. ACRL/Choice Publisher. Retrieved from https://www.choice360.org/research/research-data-services-in-academic-libraries-where-are-we-today/.

CHAPTER

5

New literacies in the age of convergence

Introduction

Although data literacy is a major issue that stands at the center of our argument, we should not forget about other literacies, because, due to their nature, it is necessary for us to take a variety of literacies into account.

As Haendel, Vasilevsky, and Wirz (2012) stated, although librarians have become experts in data management, they are an underrecognized resource in the fields of information and data literacies. The topics and skills related to these fields have always been of interest to academic librarians, and it is natural that their role is expanding to include data curation. Despite the fact that there seem to be positive developments in the recognition of academic librarians' input since this problem was first noted, much remains to be done in the provision of services and awareness raising.

This book is devoted to informing not only the researchers of today but the researchers of the future as well. The latter will primarily emerge from the ranks of graduates and doctoral students. We hope accordingly that all of them will be interested in the issues related to "new literacies," broadly meaning the development of new social practices with new technologies (Leu, Kinzer, Coiro, Castek, & Henry, 2017). When dealing with these literacies, we do not want to forget about information professionals' practical needs.

Approaching literacies

We believe that approaching literacies by repeating the question "Where's Walden?" asked by Hawkley (2017) offers an appropriate, although somewhat unusual, metaphoric view on literacies.

Research Data Management and Data Literacies
https://doi.org/10.1016/B978-0-12-824475-3.00005-9

Walden; or, Life in the Woods is the title of the book written by the 19th century American naturalist, poet, and philosopher, Henry David Thoreau (Thoreau, 1995). In this book, he reflected upon simple living in natural surroundings. However, one of the important messages goes beyond this and can be distilled from this work as follows: "Those who are able and motivated to gather information and think deeply about it have always had special access to Walden" (Hawkley, 2017, p. 6). Hawkley draws the conclusion that Walden was not the perfect search environment, but Thoreau made it a lived experience for himself, and today it is part of our job to make it attainable.

Many new models of information literacy (IL) share constructivist views on teaching and learning, arguing that "learning is an experiential and empowering process involving the continuous building, amending, and eventually the transforming of previous knowledge structures" (Walton & Cleland, 2017, p. 583). The importance of such views will be highlighted in Chapter 6, titled *Information and data professionals: Roles and education.*

The second part of this chapter's title ("the age of convergence") is an expression that appears in a number of writings with the same title. However, we relate it to Dan Ramsden's book, in which he describes how borders separating disparate fields fade due to changes in media and technology (Ramsden, 2013).

Although this book emphasizes research data management, the preceding phrases, i.e., "Where's Walden?" and "the age of convergence," could possibly be connected to express that the landscapes of scientific research and the workplace are not as far from each other as we would think. In the case of researchers, these contexts are not opposites, but rather complementary, although we understand that IL manifests itself in different ways in different contexts. Landscapes themselves are "the product of social and material settings and reflect the modalities of information (the agreed-upon sources) that people draw upon in the performance of their practices in working or everyday life, and therefore constitute the intersubjective agreement that informs their situated reality" (Walton & Cleland, 2017, p. 584). Nonetheless, the well-aimed and careful choice of sources is only one important part of this context, because—while rational and objective knowledge is valued by researchers—they also make use of experiential knowledge.

From information literacy to data literacy

Data literacy has its roots in information literacy (IL) in many regards, since the latter has a relatively long history and varied approaches to it have been extensively discussed in the literature. In this book, however, only some of the selected issues of IL are considered, insofar as they may help to perceive the nature of data literacy correctly. Comparing data literacy to information literacy involves reflection on the actual characteristics and features of the information environment that have influenced and determined that environment. In addition to this, the concept of data literacy itself "appeared partly as the result of a growing heterogeneity and complexity of information, information resources, and information structures" (Koltay, Špiranec, & Karvalics, 2016, p. 127).

The results of a bibliometric study by Dora and Kumar (2019) show that varied aspects of information literacy appear in the literature with high frequency. The connection between data literacy and information literacy was evidenced in the bibliometric study of Zhang and Eichmann-Kalwara (2019).

When examining the evolution of terms from 1975 to 2018, Onyancha (2020) found that the most frequently used ones were IL, digital literacy, and media literacy. He also found that data literacy occupied the 22nd place among the top 40 occurrences of literacies. Apparently, this difference has its root in the simple fact that the concept of data literacy appeared much later than information literacy and media literacy.

When investigating the evolution and diffusion of information literacy, Li, Chen, and Wang (2021) assumed that its changes could be informative for researchers and provide a better understanding of the corresponding domains. They identified several global topics. The first and most extensive one focused on students' information literacy learning, which has been the original objective of the development of IL. The second-ranked topic concerned library services for information literacy. The third topic, still occupying a relatively large share of papers, considered "information literacy practice from a broader perspective and focuses on the objective of keeping pace with the current technology, such as digital technology" (p. 11).

In regard to the development of varied literacies, we can say with confidence that it has been characterized by convergences, the roots of which are in the mergers of different information and communication technologies (ICTs) and varied forms of media (Livingstone, van Couvering, & Thumin, 2008) that—in its turn—induced convergences between varied literacies.

Information literacy and data literacy: Allies or competitors?

Several definitions of IL, including early ones, see it as associated with data. For instance, the Australian and New Zealand Information Literacy Framework says that information-literate persons obtain, store, and disseminate not only text but data as well (Bundy, 2004). Andretta, Pope, and Walton (2008) identified presenting, evaluating, and interpreting qualitative and quantitative data as a learning outcome of IL. According to Hunt (2005), data literacy education should borrow heavily from information literacy education, even if the domain of data literacy is more fragmented than the field of information literacy. Andretta et al. (2008) declared that presenting, evaluating, and interpreting qualitative and quantitative data are learning outcomes of information literacy (Andretta et al., 2008).

Both the SCONUL Seven Pillars information literacy model (2011) and the information literacy lens of the Vitae Researcher Development Framework (2012) stress that locating and accessing the information and data needed by researchers could result in producing appropriate data. Dealing with research data is important, because IL not only encompasses published information, but handles the underlying data as well. This is in accordance with one of the broader interpretations of information literacy, which recognizes that the concept of information includes research data (RIN, 2011). Similarly, Carlson, Fosmire, Miller, and Sapp Nelson (2011) underlined that expanding the scope of information literacy to include data management and curation is a logical development.

Schneider (2013, p. 135) defined data literacy as a component of information literacy, emphasizing that the boundaries between *information* in information literacy and *data* in data literacy are blurring. He sees the reason for this in the fact that these boundaries never have been rigid, as information literacy "has always been interested in the proper understanding and use of data that—only through the ability to use it—is converted into information." Si, Zhuang, Xing, and Guo (2013) stated that data-related services should be supported by professionals with excellent information literacy skills.

Calzada Prado and Marzal (2013, p. 126) posit that information literacy and data literacy are a part of a "scientific-investigative educational continuum, a gradual process of education that begins in school, is perfected and becomes specialized in higher education and becomes part of lifelong learning."

When suggesting a new framework for data literacy education, Maybee and Zilinski (2015, p. 1) also pointed out that "approaches to data literacy have typically been informed by information literacy."

The *Information Literacy Competency Standards for Higher Education* (ACRL, 2000) indicate the following skills:

- Determining when data is needed,
- Accessing data sources appropriate to the information needed,
- Recognizing source data value, types, and formats,
- Critically assessing data and their sources,
- Knowing how to select and synthesize data and combine them with other information sources and prior knowledge,

- Using data ethically,
- Applying results to learning, decision-making, or problem-solving.

These standards have been superseded by the Information Literacy Framework for Higher Education (ACRL, 2016). Nonetheless, this connection between information- and data-related skills remains valid. When recognizing IL as information behavior to obtain information through any medium, Johnston and Webber (2003) took an important step toward recognizing the previously mentioned convergences between different literacies.

The new definition of information literacy by CILIP (the Chartered Institute of Library and Information Professionals) undoubtedly helps to identify and clarify data literacy's nature and place among literacies by stating: "Information literacy relates to information in all its forms: not just print, but also digital content, data, images and the spoken word. Information literacy is associated and overlaps with other literacies, including specifically digital literacy, academic literacy and media literacy. It is not a stand-alone concept, and is aligned with other areas of knowledge and understanding" (CILIP, 2018). This definition not only recognizes convergence and puts emphasis on the inclusive nature of IL, but also indirectly acknowledges its overarching nature that has already been present in the thinking about literacies.

These ideas were present—for instance—in the SCONUL model, mentioned earlier, because it defined information literacy as "an umbrella term which encompasses concepts such as digital, visual and media literacies, academic literacy, information handling, information skills, data curation, and data management" (SCONUL, 2011).

The skills needed for successfully managing research data are tied unmistakably to both information literacy and data literacy. According to Davies, Fidler, and Gorbis (2011), the ideal researcher is able to filter information based on its importance using a variety of tools and techniques. Doing this is unimaginable without sense making, to be discussed in detail in a later part of this chapter. Since there should be no serious research without the ability to determine the deeper meaning of what is being expressed at face value, sense making is crucial. There is also a need for coupling the ability of translating vast amounts of data into abstract concepts, as well as understanding data-based reasoning. We should not forget either that data literacy in research settings is tied not only to research and to varied literacies, but involves self-management, innovative thinking, and problem-solving abilities (Lee, 2013). On a different but also general level, we can say that data literacy is not only cognate with information literacy in general, but is compatible with the information literacy focus of academic librarianship (Cox, 2018).

Acknowledging the diversity of literacies calls attention to the fact that a number of them can be recognized as overarching ones. Digital literacy, which took 2nd place among the top 40 types of literacies in the literature for 1975–2018 as identified by Onyancha (2020), is usually seen as an overarching, or broad, umbrella term (Dowell, 2019). Another literacy appearing in Onyancha's list with a lower frequency, and usually regarded to be overarching, is transliteracy, which was originally coined to support a cross-sectional approach to communication platforms (Koltay et al., 2016). A relatively recent definition of data literacy also underlines that information literacy can be understood as "an overarching, receptive competence as it refers to the handling of all forms of information including data" (Schüller, Busch, & Hindinger, 2019, p. 24).

This connection is evidenced in the bibliometric investigation of Zhang and Eichmann-Kalwara (2019), who have found the *data literacy* cluster related to *information literacy*. There is a clear difference between the high frequency of information literacy's appearance contrasted to the 22nd place of data literacy among the top 40 occurrences of literacies (Onyancha, 2020). Apparently, this difference is caused by the simple fact that the concept of data literacy appeared much later than information literacy and media literacy.

Concerning disciplinary affiliation, it was found that across 26 unique Scopus subject areas, IL maps onto social sciences with hits in 76.88% of the articles. Nonetheless, it is no longer restricted to social sciences, but is spread across 27 disciplines. There is no definite explanation for this development, but "it is apparent that information literacy has evolved over time and we are likely to witness additional changes in terms of the following: learning and teaching, practice, instruction, technological uptake, relationship with other literacies, disciplinarity, content and context and so on" (p. 119).

The relationship to media literacy and MIL

Data literacy "shares some distinguishing features with media literacy, where the use of tools to use and reuse content in ways not imagined by the content creator are a critical part of this literacy" (ACRL, 2013, p. 11). This idea supplements the view on information, according to which "messages, as media expressions, are part of the broader term of information that includes media constructs as well as any other encapsulated knowledge or data in any format/medium including printed, electronic, digital, images, sound, or realia" (Lau, 2013, p. 78).

As in their bibliometric study, Lee and So (2014) found information literacy and media literacy share the same or similar goals. Nonetheless, it is clear that information literacy has a clear focus on library and information science and technology, while media literacy's scope is broader, as it is more closely related to communication, health-related issues, leisure, and culture.

Although data may be produced in seemingly objective processes, it is not inherently objective. Data literacy involves asking questions that are similar to those asked when practicing information literacy, such as inquiring about the creator(s), the ways of creating the data, and identifying the presence of biases. Certainly, we should be aware of a possible disparity between the provenance of data and information. While human biases may characterize both, machine-based biases are present only in datasets, because information is produced and disseminated by humans only. Furthermore, we have to be aware of the fact that data can be created without direct input, while new information should be based on previous input (Fontichiaro & Oehrli, 2016).

Abner (2020, p. 20) characterizes the situation as follows: "A data literate person will ask questions about how the value is determined, who assigns the value, and what societal factors might influence the value. These are research questions in their own right, and the answers are not always readily (if at all) available. However, when learners understand the process of data creation as wholly guided by humans, they think critically about that process and how that affects any data they examine. These effects can be mitigated by supplementing a data-based result with other pieces of information."

By stressing the need for critical approaches, data literacy can be subsumed within a core skillset of being able to *discriminate* or exercise discernment. Discernment is being able to make independent judgments on the validity of information and evaluating the information component of literacies (Mason, Khan, & Smith, 2016; Walton, Pickard, & Dodd, 2018). This substantiates arguments that the final goal of data literacy is fostering a culture of the semantic scholar, because—unless educated in data literacy and information literacy—scientific training alone does not guarantee that researchers become aware that their own scholarly communications may constitute primary sources of data (Haendel et al., 2012). This type of data literacy should involve abilities to translate vast amounts of data into abstract concepts, to understand data-based reasoning that leads to seeing almost everything that we come into contact with through the lens of data, and will regard it as computational and programmable.

Data must be understood and interpreted. These crucial activities may be aided by deconstructing messages. This need also provides a proof of convergence between media literacy and data literacy. Deconstruction in media environments is described by Potter (2019), among others. Obviously, deconstructing texts and media messages are in many regards different from data. Nevertheless, deconstruction is apparently feasible if we examine the epistemological status of the data, because we can break down and reinterpret the statements on which the datasets are based if we want to attribute professional authority to them.

Breaking down messages into meaningful elements and then judging the value of these elements by comparing them to well-defined criteria can be useful for information literacy tasks. This can be repeated—mutatis mutandis—in the case of datasets. Other skills, such as determining which elements are alike or different in some ways, followed by making inferences based on patterns discoverable in a small set of elements, and then generalizing the pattern to all elements in the set, are also useful. With all things considered, deconstruction promises to play a role in ascertaining the objectivity of data.

As Gebre and Morales (2020, p. 1) wittily observed, "Datasets do not speak for themselves, because they require context for analysis and interpretation." This seemingly simple, commonplace sentence points out that, no matter what data we speak about, it can represent scholarly research but its use is limited due to the lack of data literacy skills, or because of the presence of a "data speaks for itself" mentality. This opens an intriguing perspective, especially if we look at the possibility of an opposite perspective. Koesten, Gregory, Groth, and Simperl (2021) published a paper titled "Talking datasets – understanding data sense-making behaviours." The first part of this title seemingly contradicts the preceding statement, because the authors see the importance of identifying and detailing "common patterns of data-centric sense-making across three clusters of activities" that they present "as a framework: inspecting data, engaging with content, and placing data within broader contexts" (p. 1). Nonetheless, sense making is a step forward to "making data to speak." Sense making could go side by side with understanding. According to Bawden and Robinson (2016, p. 297), the "impacts of developing understanding are likely to be diffuse and long term, but they are also likely to be highly significant, since a change in understanding is likely to provide changes to a much greater extent than is simply acquiring information." As a higher-order knowledge, understanding has a coherent and explanatory potential. Although often neglected as an additional layer in the well-known Data-Information-Knowledge-Wisdom (DIKW) hierarchy (mentioned in Chapter 1, Information, data, text, document), it is a kind of higher-order knowledge, having explanatory power and dealing with specific and isolated information and documents. In our opinion, although understanding cannot be directly tied to literacies, it can at least supplement sense making.

Consciousness is also an important related construct, which is not the same as knowledge, but is necessary for knowledge and knowledge growth. As Budd (2020) posits it, the development of consciousness and conscious states is essential to the efficacy of IL. Meaning is discerned through interpretation, but it "does not necessarily equate to accuracy or truth, or even ethical communication. Determining those elements is a vital component of the conscious process of interpretation. When it comes to such things as academic exercises, the determination of such things as accuracy is a necessity (not merely for academic success, but for knowledge growth). In other words, the entirety of IL affects the entirety of consciousness" (p. 1389). Apparently, we can extend consciousness to data literacy as well.

The evolution of information literacy and media literacy and the resulting overlap between them apparently fosters the same skills while addressing different information constructs (Lau, 2013). This was one of the motives for framing media and information literacy (MIL), which combines the values of information literacy with those of media literacy. The definition of media by Robinson (2009, p. 584) as "any information-bearing entity type expressed specifically or generally: for example, books, digital information, photographs, the web, museum objects, etc." indirectly verifies the correctness of this endeavor.

There are several other arguments as to why MIL should be treated as an entity on its own right, due to the different emphases of information literacy and media literacy that have remained palpable. On the other hand, contemporary media education and information literacy education cannot be separated from each other, at least not in the perception of their users. The main reason for this is that "the experience of being a user of information resources and a consumer of media is so similar that most people would not differentiate between the two so having two separate forms of education related seems peculiar. Media and information texts do not exist in discrete 'silos'; access to them is afforded through integrated platforms" (Leaning, 2013).

Although this book puts less weight on participatory approaches, it should be mentioned that MIL attaches importance to them, and thus they cannot be left out completely.

As, for instance, Wilson (2012) points out, the objective of MIL is to broaden civic participation in the media, as related competencies enable individuals not only to access, organize, and evaluate information and media messages, but to produce such messages, although each medium or information source enables creating meaning differently. The eminence of production is also indicated in the *Intersections of Scholarly Communication and Information Literacy* report of the Association of College and Research Libraries, which also declares that the comprehension of all kinds of content, including text and data, is a critical outcome of MIL (ACRL, 2013).

The relationship to academic literacy and statistical literacy

The strength of the connection of data literacy to these literacies is different. Data literacy and academic literacy are tied together rather loosely, but have the potential to weave their bonds tighter. Statistical literacy is often mentioned alongside data literacy. Furthermore, in some cases it is treated as a constituent of data literacy.

Nonetheless, the relationship between data literacy and academic literacy has many faces. The latter concept and its theoretical background are not new, but have not lost their actuality.

The best-known aspect of academic literacy is embodied in writing activities, so it is close to information literacy. However, academic literacy education should "expand across domains, as well as to new information modes and formats" (Norgaard & Sinkinson, 2016, p. 27).

Academic writing "strives to draw accurate conclusions based on evidence." However, claims may be rooted not only in textual analysis, but drawing conclusions is often based on looking at data, then analyzing and presenting it, as well as evaluating and executing and improving it (Shome, 2019, p. 2). In a more general context, it can be said that—from the point of view of this book—the emphasis of academic literacy education should not be limited to providing help to students for "surviving" at the university by producing correct written documents, but to developing their potential for becoming researchers in the future also through producing and analyzing research data.

Statistical literacy, defined as "the ability to read and interpret summary statistics in the everyday media: in graphs, tables, statements, surveys and studies" (Schield, 2010, p. 135) is often mentioned in connection with data literacy, due to the existence of a common set of problems and a similar level of approach between the two literacies.

As put by Schield (2004, pp. 6–8), the relationship between these two literacies is based on sharing the need for evaluating information. He also posits that information literacy, statistical literacy, and data literacy are interrelated; thus "it is difficult to promote information literacy or data literacy without promoting statistical literacy."

He indicates the reason for this as follows:

- Statistical literacy studies the use of statistics as evidence in arguments,
- A substantial part of information involves statistics, and statistical data is a frequent form of information's appearance,
- The key to being statistically literate is that all statistics are socially constructed, because people "choose what to count or measure, how to assemble those measurements into summary statistics, what comparisons to form from these statistics and how to communicate these statistics."

Statistically literate students must be able to "think critically about basic descriptive statistics, analysing, interpreting, and evaluating statistics as evidence." In contrast to this, data-literate students are required to "be able to access, assess, manipulate, summarize, and present data."

In this way, Schield creates a hierarchy, where critical thinking is on the upper level by involving analysis, interpretation, and evaluation. Beneath critical thinking, we find information literacy, then statistical literacy. Data literacy occupies the bottom place.

As he sees it, information literacy requires both statistical literacy and data literacy, and statistical literacy is an essential component of information literacy. On the other hand, data literacy is an essential component of both information literacy and statistical literacy. Both information literacy and data literacy should be expanded to include critical thinking and statistical literacy. Data literacy involves training individuals to access, assess, manipulate, summarize, and present data, whereas statistical literacy aims to teach how to think critically about descriptive statistics.

Stephenson and Schifter Caravello (2007) underline the importance of data and statistical literacies as components of information literacy in the social sciences, arguing that the ability to evaluate information essentially requires that one understand the data and statistics used in an information resource. According to Calzada Prado and Marzal (2013), this means that data literacy is equated to statistical literacy.

As Hunt (2005, p. 14) points out: "in the practical implementation (or on the ground) data literacy is quite different than traditional information literacy." The origin of this difference lies mainly in the greater complexity involved in handling data than any other type of information (Thompson & Edelstein, 2004). Nonetheless, their similarities may lead to regarding data literacy as a complement to or a form of information literacy, as respectively suggested by Stephenson and Schifter Caravello (2007), which makes us think that data literacy would be the umbrella concept covering statistical literacy, rather than vice versa.

The earlier discussions about the relationship between data literacy and statistical literacy continued in the second half of the 2010s. Mason et al. (2016) stated that data literacy involves understanding the need for collecting data and the various ways of doing it, and taking into consideration that—beyond basic literacy—there is a need for being numerate, i.e., properly understanding formulas, graphs, charts, and tables. The difficulties of naming literacies can increase. As Macy and Coates (2016, p. 314) put it, a possible perspective places data information literacy "under the broad umbrella of digital literacies sharing common elements with numeric and statistical literacy."

Definitions and the nature of data literacy

The definitions of data literacy, the majority of which were conceived in the 2010s, are varied. In some cases, not only the definition contents differed but the literacies' names as well.

Somewhat earlier than most definitions, in 2008 the expression *information literacy in the data world* appeared (Fosmire & Miller, 2008). Two years later, data literacy was defined plainly as the ability to understand, use, and manage data (Qin and D'Ignazio (2010). Although under a somewhat different name, i.e., *research data literacy*, a relatively simple and broad definition says that it is a "human competence to locate, analyse, organize, present, and evaluate research data" (Schneider, 2013, p. 136).

Data literacy, as it is understood by the Association of College and Research Libraries, focuses on understanding how to find and evaluate data, giving emphasis to the version of the given dataset and the person responsible for it, and does not neglect the questions of citing and ethical use of data (ACRL, 2013).

As stated by Calzada Prado and Marzal (2013), data literacy enables individuals to access, interpret, critically assess, manage, handle, and ethically use data. They identified data literacy as the ability to understand and use data effectively to inform decisions applying data literacy skills, which include knowing how to identify, collect, organize, analyze, summarize, and prioritize data. They also included the issues of developing hypotheses, identifying problems, interpreting the data, and determining, planning, implementing and monitoring courses of action.

Synthesizing several existing definitions, Ridsdale et al. (2015) set up a simple definition, according to which data literacy "is the ability to collect, manage, evaluate, and apply data, in a critical manner." (p. 45) They add that data literacy, and the related set of learning outcomes are on the way to becoming recognized as a necessary skill set, because data literacy encourages cross-disciplinary thinking, helping break the academic silos and enabling creative and critical thinking.

A more detailed, composite definition describes data literacy by following the line of thought laid down in some of the previous publications. It says that data literacy is "a specific skill set and knowledge base, which empowers individuals to transform data into information and into actionable knowledge by enabling them to access, interpret, critically assess, manage, and ethically use data" (Koltay, 2017, p. 10).

In a recent definition, Husfeldt and Alt (2021, p. 75) state that data literacy "describes the handling of data for the purpose of transforming data into knowledge. It requires an understanding of the possible reception as well as an understanding of the data-generating process, e.g., the origin of data, its limitations, and the analysis tools used."

Discussing the nature of data literacy is in many regards a continuation of defining the previous terms and extending them. For instance, when looking at the possible context of data-related academic literacy, we may take several goals of "traditional" academic literacy, then substitute *data* for *information*, which may help in identifying a number of requirements that are relevant for data literacy.

Data literacy education (to which we will return in the Chapter 6, *Roles and education of information and data professionals*) frequently has two partially overlapping purposes:

- Producing students and researchers who are data-literate, successful researchers.
- Educating data management professionals (Qin & D'Ignazio, 2010; Schneider, 2013).

Although data literacy is on its way to becoming a standard subject in graduate and doctoral education, the majority of researchers are acquiring data literacy in the workplace. As Ince et al. (2018, p. 167) put it, "while the challenge of managing literature is an early and obvious one in a doctoral student's career, an additional important aspect of KM is data literacy and data management."

As stated previously, data literacy began to gain momentum only in the 2010s, as the capacity to store massive amounts of data and forward them on high-bandwidth networks moved data to the forefront in practically all sectors of society. This development has increased the attention toward research data, data in the business world, and in the public sphere. As a result, the need for data literacy appeared (Koltay, 2015). The ideas characterizing the thinking about data literacy are often attached to active and informed citizenship. It can be supposed that the concept of data literacy emerged not exactly because of the convergence between different literacies, but these convergences revealed literacies' potential to broaden their circle to a newer medium, not to mention the interest in data among members of varied academic communities.

The terminology in this field is not entirely standardized in the sense that there is *science data literacy* (Qin & D'Ignazio, 2010), *data information literacy* (Carlson et al., 2011), and research data literacy (Schneider, 2013). Including the word *science* into some of the names reflects a relative primacy of different kinds of data in research settings. Beside its apparently high frequency, in this book we voted for the term *data literacy*, because it is simple and straightforward (Koltay, 2015).

Describing the nature of data literacy would be inconceivable without examining the conceptual similarities and differences between information and data that have been treated in Chapter 1, *Information, data, text, and document*. However, there seems to be no need to repeat the definitions of these fundamental concepts. Having said this, it should be emphasized instead that data literacy extends the majority of ideas and rules, laid down in the studies, addressing information literacy, media literacy, digital literacy, and other literacies. Active publications on data literacy seem to have characterized the period between 2005 and 2015 and, not surprisingly, the *data literacy* cluster in Scopus was found to be related to the cluster of information literacy (Zhang & Eichmann-Kalwara, 2019).

According to Tewell (2015), data literacy can also be interpreted from the standpoint of its inherent critical qualities, which are also highlighted in the composite term *data information literacy* used, e.g., by Carlson et al.

(2011). The latter regard data literacy as part of information literacy and a kind of logical development of information literacy, and interpret it from the standpoint of its inherent critical qualities. Just like in the case of information literacy, criticality and critical assessment are considered as essential features of data literacy.

As Špiranec, Kos, and George (2019) see it, the discussions around data literacy can be seen from several perspectives, including the need for reimagining literacy concepts. They posit that the ontological treatment of data can be subsumed under contextuality, interpretativeness, and the transparency of data. They argue that defining data literacy as the ability to understand, find, read, interpret, evaluate, manage, and use data (e.g., Koltay, 2017) "does not reflect the holistic approach to literacies, as seen in more recent conceptualizations of literacies commonly discussed in the field, e.g. information literacy, metaliteracy, media literacy." This means that approaches to these literacies approximate data literacy to critical information literacy.

As Pangrazio and Sefton-Green (2020, p. 6) summarize it, "Operational definitions are concerned with what might constitute acceptable functioning literacy, whereas critical literacy brings with it a host of 'higher order' questions, including the ability to challenge, ask questions and reflect, rather than simply decode symbols in the way that a machine might. When applied to data, it is very difficult to understand what the equivalent of operational literacy might be without a critical dimension. Probabilistic and statistical reasoning (which by definition will rely on confident use of various mathematical processes) underpin thinking about how data works at both a functional and a critical level."

Kitchin (2014, p. 3) warns us that big data ushers in "a new era of empiricism, wherein the volume of data, accompanied by techniques that can reveal their inherent truth, enables data to speak for themselves free of theory. The empiricist view has gained credence outside of the academy, especially within business circles, but its ideas have also taken root in the new field of data science and other sciences. In contrast, a new mode of data-driven science is emerging within traditional disciplines in the academy" (p. 3).

In contrast to new forms of empiricism, there is data-driven science, which remains true to the scientific method, as we know it. At the same time, it is willing to generate hypotheses not from the theory, but from the data.

Data literacy covers "both the creation of data products by the methodically experienced specialist as well as the competent handling of data by the end user, i.e. the critical and adequate interpretation and application of the data" (Schüller, 2020, p. 6).

Criticality is undoubtedly important, but the crux is that it should influence education on data literacy. In addition to this, focusing on the citizen instead of on the professionals, in particular on the researchers, receives less emphasis, when we discuss data literacy against the background of research data management, as is done in this book, although there is no strict line of division between them. Even if looking at data literacy from the point of view of the researchers, data literacy can have a true lifelong-learning character, or—in some measure—it should evolve in this direction. The reason for this is that researchers in the past were not required to possess many of the data-related skills. At least, the skills related to data sharing and the need for citing data were not emphasized, not to mention the initiatives, such as the FAIR principles, requiring attention to Findability, Accessibility, Interoperability, and Reuse, as described in Chapter 2, Information and data ecologies.

The expression *digital literacy* also frequently appears in the literature. In accordance with this, it is the 10th most frequent concept in the list, as compiled by Onyancha (2020). Notwithstanding, discussing it does not seem to bring any additional value to our present discussion, because it is usually seen either as a literacy on its own right, or as one of the literacies incorporated into information literacy.

To sum up this part of our argument, we can say that data literacy can be seen from different points of view, such as being related to issues of the civil society, open government, community informatics, journalism, business, and teacher education. It can be approached also as a transversal competence, related to the "research skill for students and professionals – accessing existing data sets to produce and communicate new knowledge, making scientific experiments robust and reproducible" (Corrall, 2019, p. 2). While appreciating a number of broad societal missions, in this book data literacy is chiefly examined in a relatively well-defined sense, i.e., narrowing its meaning down to the relationship to scholarly research, and in particular to research data.

According to Pangrazio and Sefton-Green (2020, p. 13), literacy is "a useful shorthand to explain how individuals are able to make sense of complex underlying patterns and to relate them to commonly shared and understood theoretical insights." They add that the most conventional definitions of literacy contain reading and writing, which are interrelated, but show two, necessarily distinct, dynamics. This fact is important for data literacy, because data literacy would require the ability to "read the signs inscribed in data." On the other hand, the ability to read data inscriptions also relates to the capacity to make them.

Data literacy processes and competencies

As Carlson and Johnston (2015) pointed out, the community of practice for data librarians differs from that of information literacy. Nonetheless, this should not lead to creating new silos.

Data literacy competencies must be applied to research-related tasks and related issues of citizenship. Nonetheless, as promised earlier, we will focus on academic purposes, thus on selected data literacy competencies that can be used in the following fields:

- Discovery and acquisition of data,
- Data management,
- Data curation,
- Data citation,
- Data analysis,
- Data conversion and interoperability,
- Data reuse,
- Data preservation,
- Data visualization,
- Ethics, and
- Metadata (Carlson et al., 2011).

Ridsdale et al. (2015) set up a matrix of data literacy competencies with the intention to foster an ongoing conversation about standards of data literacy and learning outcomes in data literacy education. The perhaps most important activity in this matrix is quality evaluation, which includes assessing sources of data for trustworthiness and for errors or problems. Evaluation appears even when we collect the data. While data management includes data organization, metadata creation and use, and data preservation, data curation is associated with security and reuse. Data interpretation clearly shows the mechanisms that also characterize information literacy. Data visualization comprises creating, evaluating, and critically assessing graphical representations of data. Besides visualization, presenting data verbally in a clear and coherent manner is also a crucial data literacy competence.

Ridsdale et al. (2015) identified data literacy as one of the research data service activities that supports researchers in building the skills and knowledge required for managing data well. Therefore, we can say that data literacy is related to practically all processes that are covered by RDSs that build the main framework of libraries' involvement in supporting the data-intensive paradigm of research (Tenopir, Sandusky, Allard, & Birch, 2014). RDSs are undoubtedly comprehensive, thus data literacy's coverage of their aspects also makes it overarching and comprehensive.

Maybee and Zilinski (2015) have found that information is often used in different disciplinary contexts. However, its discipline-specific nature has been apparently examined concerning information literacy. Obviously, we can take into account the close ties between IL and data literacy, discussed previously.

In a bibliometric study, Pinto, Pulgarín, and Escalona (2014) observed that disciplinarity influenced information literacy both in the health sciences and in the social sciences due to their specific personalities.

Bawden and Robinson (2017) examined the history of chemical information literacy from the beginning, i.e., considering it also for the periods before computer searching and digital resources. They found that—while chemical information literacy contains some generic elements—it is more strongly domain specific than any other subject. By doing this, they reflected on "a continuing debate on whether there can be such a thing as generic information literacy, or whether each discipline or domain must have its own variant, with information literacy taking on a different 'personality' in different domains, with different concepts and relationships" (p. 19).

As Farrell and Badke (2015) underlined it, in order to meet the demand of the information age for skilled handlers of information, literacy education must become situated within the sociocultural practices of the disciplines by an expanded focus on epistemology and metanarratives. Truly situated IL will therefore require that librarians or teaching staff members invite students into disciplines with the understanding that the information practices belong to a discipline. In a somewhat similar vein, Calzada Prado and Marzal (2013) emphasize the ability of identifying the context in which data are produced and reused.

It is rather obvious that (despite criticisms of information literacy for the same reason) data literacy can and should be approached by identifying and developing skills and abilities.

As Carlson, Fosmire, Miller, & Sapp Nelson (2011, p. 5) put it, "data literacy involves understanding what data mean, including how to read graphs and charts appropriately, draw correct conclusions from data, and recognize when data are being used in misleading or inappropriate ways."

Knowledge areas, identified by Ridsdale et al. (2015) were as follows:

- Data management, including data organization; data manipulation; data conversion; metadata creation and use; data curation, security, and reuse;
- Data evaluation, including analysis and interpretation, identifying problems of using data, and data visualization;
- Data application, including critical thinking; data culture; data ethics, data citation; data sharing, evaluating decisions based on data.

Based on the examination of data literacy definitions, Fontichiaro and Johnston (2020, p. 76) have found that data literacy "incorporates numeracy, quantitative literacy, and mathematical and statistical calculations, as well as problem solving, communication, and decision-making."

Focusing on data management and data science, Grillenberger and Romeike (2018) devised a model of data literacy competency. They emphasize that data management focuses on rather static aspects of storing and accessing data appropriately, while data science sets its focus on data analysis and visualization, which are dynamic aspects. Although devised for secondary school teachers of computer science, this model promises useful insights into the theory of data literacy and—at the same time—offers guidance for the education of information professionals by naming the main relevant areas working with research data:

- Data analysis,
- Machine learning,
- Big data,
- Data privacy and data ethics,
- Data storage,
- Core technologies/systems for storing and managing data design principles (ways for accessing data, requirements on data stores and data storage).

Grillenberger and Romeike also considered the data life cycle (also addressed in Chapter 4, "Research Data Management"), resulting in additional insights and identification of 11 phases:

- Acquisition,
- Cleansing,
- Modeling,
- Implementation,
- Optimization,
- Analysis,
- Visualization,
- Evaluation,
- Sharing,
- Archiving, and
- Erasure.

Computational thinking (CT) takes a growingly significant place among the competencies of data literacy.

As research is becoming largely determined by computing, the previously described abilities and skills required for data literacy begin to fit into the framework of computational thinking that might be incorporated into the general thinking about scholarly research, and that results in changes in its ontologies and epistemologies.

Wing (2006, pp. 33–35) defines CT as "a way of solving problems, designing systems, and understanding human behaviour that draws on concepts fundamental to computer science. It represents a universally applicable attitude and skill set everyone, not just computer scientists, would be eager to learn and use." She also adds that it "requires thinking at multiple levels of abstraction, as it is a fundamental skill that every human being must know to function in modern society" (p. 35). While using fundamental concepts of computer science, CT focuses on thought processes that can be operationalized (Wing, 2008).

As early as 2007, Bundy stated that "computational thinking is influencing research in nearly all disciplines, both in the sciences and the humanities" (Bundy, 2007, p. 67). In our opinion, CT can influence not only scholarly disciplines, but is required for successful information and data literacy activities. Relating computational thinking to data science, Song and Zhu (2017) see it as a basic problem-solving framework. In the inventory of core skills required for librarians for data science, Federer, Clarke, and Zaringhalam (2020) identified CT as an important skill.

Computational thinking skills include algorithmic thinking and navigating multiple levels of abstraction and data modeling, taught with or without the use of computers (Jacob & Warschauer, 2018, p. 3).

Computational thinking involves understanding how to break a problem apart into workflows, which can also be an important approach to problem solving for librarians working in DS/OS. However, its effect goes beyond this, because CT is a general method of human problem solving, which might become one of the general approaches to scholarly research, as digital devices require objects to be translated into digital code and then "internally transformed, depending on a number of interventions, processes or filters" (Berry, 2011, p. 1).

Despite its importance, data-literate persons must also be aware of the limitations of computational thinking. This requires that they understand the limitations of data and remain able to act in the absence of data (Davies et al., 2011).

A framework for data literacy

Although we want to concentrate on data literacy, information literacy also comes into the picture. Its history is much longer than that of data literacy. Their kinship has already been explained in this chapter, but there is more to say about the relationship between the two.

As Townsend, Brunetti, and Hofer (2011) have put it, information literacy program statements and policies generally espouse goals of critical thinking and lifelong learning. However, information literacy often continues to be taught as a kind of enhanced bibliographic instruction. This phenomenon is probably rooted in the difficulty of relating theoretic IL to practice.

The Information Literacy Competency Standards for Higher Education (ACRL, 2000), conceived by the Association of College and Research Libraries, are an example of approaching information literacy as a set of measurable skills. It also served for a long time as the perhaps most influential document in this field. However, it has been rescinded and replaced by the Information Literacy Framework for Higher Education (from now on the *Framework*) (ACRL, 2016). The Framework presents a shift away from "standards, learning outcomes, or any prescriptive enumeration of skills" and toward "a cluster of interconnected core concepts, with flexible options for implementation" (p. 2). It contains practices and dispositions related to required proficiencies and capabilities. However, unlike the Standards, it does not identify learning outcomes, but lets individual institutions transform approaches, articulated in the form of threshold concepts, i.e., ideas, forming passageways to better understanding or ways of thinking and practicing within a given discipline into learning outcomes that best fit student needs.

Threshold concepts are basic ideas and processes that define a field of science, but are so ingrained in the given field that they often remain unspoken and unknown to practitioners. Threshold concepts prompt us to change our approach. They are also integrative, as they combine separate concepts and help to define the extent and boundaries of a (sometimes unique) field of science (Hofer, Townsend, & Brunetti, 2012).

This is perhaps the reason to say that—while we can agree with many ideas set forth in the Framework—it is more difficult to do so in relation to the underlying threshold concepts. These are abstract concepts that form the basis of a discipline, but it is often not clear to many instructors what is meant by threshold concepts (Fister, 2017).

Threshold concepts represent information previously not known by the learner, and once a person learns and understands that information, it permanently and drastically changes their perception of the topic (Meyer, Land, & Baillie, 2010).

The Framework formulates the following threshold concepts:

- Authority is constructed and contextual;
- Information creation as a process;
- Information has value;
- Research as inquiry;
- Scholarship as conversation;
- Searching as strategic exploration.

Although in our opinion it would have been more advantageous to form some of these concepts in declarative sentences (e.g., "Information creation is a process"), the ideas that they cover deserve not only the attention of IL professionals, but are useful for media literacy professionals as well. Obviously, such thinking would require that we change *information* to *media*, and properly adjust these concepts to the goals of media literacy education.

Recently, a new and notable development appeared, thanks to the ideas of Abner (2020). She developed a preliminary core conceptual framework for data literacy. One of the premises of her venture was that information literacy for the digital humanities should embrace many literacies, but their starting point must be data literacy. Her proposal is seemingly limited in scope, as she did not expressly address researchers beyond the digital humanities. Nonetheless, her message deserves the interest of a wider audience, including different information and data professionals. By the same token, any researcher could also give attention to it, because we must see that three threshold concepts enumerated in the ACRL Framework seem to be outspokenly relevant to research and researchers. The related concepts are:

- Research as inquiry;
- Scholarship as conversation;
- Searching as strategic exploration.

The first threshold concept in the data literacy framework conceived by Abner emphasizes that data is crated both by machines and by humans. The general explanation added to this is that "data is not inherently objective, even when created through a seemingly objective process" (p. 20). In this regard, three key questions have to be explored:

- Who or what created (or guided the creation of data)?
- How was the data created?
- What human or machine-based biases are present?

As Abner sees it, most definitions of data literacy can be distilled to the ability to critically create, manipulate, manage, analyze, understand, and communicate data. She adds that the difference between IL and data literacy is that the former typically covers forms of knowledge directly created by humans, while producing data may not need direct input.

Data literacy in research and for civic life

The work of today's researchers mobilizes competencies on three levels:

- Conceptual competencies that include, among others, innovative thinking, problem solving, and critical thinking,
- Human competencies, like social networking skills, self-management, and cross-cultural interaction skills, and
- Practical competencies that include media literacy and information literacy (Lee, 2013).

Unfortunately, it would be utterly naïve to presume that researchers would readily accept the need for acquiring data literacy skills. It is enough to look at evidence that not only people in general, but also researchers, hold themselves competent and skillful in dealing with information (Nicholas et al., 2008). Most probably, data occupies a similar place in researchers' perceptions. Many researchers have been working with data extensively and over a long time, often much before the importance of research data became known to wider circles. This suggests that they may feel confident with managing varied datasets. On the other hand, we should not forget that—as mentioned in Chapter 4, "Research Data Management"—data was not usually recognized by them as something separated from other tools and processes of research (Ojanen, Lindholm, & Siipilehto, 2020).

In the data literacy framework elaborated by Abner (2020), data is seen as part of the scholarly conversation, because it is an artifact of the iterative process of research; thus we should ask what kind of data is being shared, and what steps are being taken to ensure that the data is made findable, reusable, and preserved.

In contrast to data literacy initiatives, which focus on developing technical, computational and statistical competencies for working with datasets, Gray, Gerlitz, and Bounegru (2018, p. 1) propose data infrastructure literacy, which "encourages critical inquiry, imagination, intervention around the infrastructures for creating, using, and sharing data." They intend to go beyond data as an informational resource, and in this way "equip people with data skills, but also to cultivate sensibilities for data sociology, data culture and data politics," adding that critically oriented views are broader than earlier ones, and often deconstruct the ideology that underpins datafication. The situation therefore is similar to the aspirations of critical information literacy, through which we can look at information literacy as something less prescriptive and measurable, aimed at answering the needs created by the prevalence of social media (Špiranec, Banek Zorica, & Kos, 2016).

Critical information literacy and critical data literacy come into the picture apparently if they are addressed to the broad public and/or used in the social sciences, often in relation to those applications of big data, which lay the foundation of mythologies about large datasets offering a higher form of intelligence (Boyd & Crawford, 2012). As there is usually big data involved, it is *critical big data literacy* that may contribute to the development of relevant views and skills (Sander, 2020), especially for digital citizens, by focusing on potentially problematic societal issues.

As portrayed by Oliphant (2017), big data is Janus-faced, at least in the attitudes regarding its impact. In some cases it is uncritically accepted and embraced as a tool that is capable of fundamentally altering practically every sector of society. On the other hand, it is emphasized that big data can be used for surveillance, breaching privacy. It can also be hacked, searched, aggregated, and preserved (often without consent). Besides detecting these threats, we see efforts to use data for the public good and to empower people. In order to support such moves, we must understand that data literacy "can affect everybody in an institution, regardless of their role in a research project, the characteristics of the group, or the type of research. Transversal services transmit generalizable knowledge that has to be interpreted for each application. It takes a long time to create these services, but once they are created, it is relatively straightforward to apply them to specific cases" (Llebot & Rempel, 2021, p. 21).

Data literacy cuts across disciplines. It involves knowledge and the ability to apply this knowledge, and a value-driven willingness to do so. At the same time, data literacy can serve various purposes. On the one hand, it is required for using data professionally in the context of a specific subject area. There should be decision-making competence that allows distinguishing between data, information, and opinion. On the other hand, data literacy can be understood as an ability necessary for responsible citizens to make informed decisions in everyday life and in various political questions (Schüller & Busch, 2019).

Data is a small part of the whole narrative, characterizing research, so it cannot give a complete picture in itself. This fact should motivate us to explore what the data means for us, looking for additional information that confirms this. Indeed, data serve as a "snapshot" representation of a particular reality, and we tend to believe data-driven conclusions more readily than we do in the case of other forms of knowledge. This statement could be coupled with the understanding that data as such is not inherently more objective than a journal article or a conference presentation (Fontichiaro & Oehrli, 2016). While we may try to remove bias from our computing processes, decisions made during the creation process influence the final dataset because "people choose what to count or measure" (Schield, 2004, p. 7). Although this message should be understood by wide circles of society beyond the academic sphere, it is also important for researchers and academic librarians. A similarly wide-reaching, transversal issue is that data-literate persons should recognize the gaps or discrepancies between data and reality. Being aware of these gaps and understanding the limitations of data-driven analysis, as well as looking for information to complete a narrative or verifying claims derived from a dataset, are highly desirable skills (Davies et al., 2011).

Competency-oriented perspectives put emphasis on developing technical and conceptual skills to enable dealing with mainly quantitative data, including determining its quality, credibility, and relevance as well as being able to analyze and interpret it. The empowerment-oriented perspective also subscribes to the competency-oriented view of data literacy, but goes beyond technical and conceptual skills. This view focuses "on the use of data literacy as a means to foster civic engagement and build equitable and democratic society" (Gebre, 2018, p. 333).

The promise of engaging and addressing community-related issues will become reality when the lack or a part of data literacy skills among actual and potential users will develop to a satisfactory level of skills. Until this, researchers and data scientists will remain almost the sole group of data users (Gebre & Morales, 2020).

Other relevant literacies

There are a large number of varied literacies. In this section, however, we will provide short overviews only of visual literacy and AI literacy.

Visual literacy

Visual literacy (VL) is a relatively new member of the literacies family, which evolved to meet the requirements and effectively exploit the potentials resulting from the development and pervasiveness of new digital technologies (Leu, Kinzer, Coiro, & Cammack, 2004). Also known as *digital visual literacy*, *visual fluency*, or *visuacy*, it focuses on "examining how people make connections between the visual texts they encounter and their own experience of reality, examining media texts for their plausibility and correspondence with other media representations" (Hobbs, 2006, p. 22).

Defined as "a set of abilities that enables an individual to effectively find, interpret, evaluate, use, and create images and visual media. Visual literacy skills equip a learner to understand and analyse the contextual, cultural, ethical, aesthetic, intellectual, and technical components involved in the production and use of visual materials. A visually literate individual is both a critical consumer of visual media and a competent contributor to a body of shared knowledge and culture." There are cases when information may be better presented visually, compared to verbal communication. Images may even substitute for text; thus the ubiquity and importance of visual media are changing the meaning of being literate in the 21st century (ACRL, 2011). This concerns any combination of body language, motion, and dance, works of art, photographs, films and videos, museum exhibits, advertisements, illustrated written or verbal discourse, architecture, hypermedia, and virtual reality experiences (Chauvin, 2003).

Visual literacy occupies the eighth place on the list of top literacies assembled by Onyancha (2020) for the years between 1991 and 2000.

As explained by Federer (2016, p. 40), data visualization is, strictly speaking, not part of the data management process. Nonetheless, it is "an important skill for working effectively with data. Visualizations are often used to convey final research results and demonstrate the researcher's findings, but data visualization may also be exploratory in nature, providing a means to visually identify patterns that might otherwise have been difficult to discern in complicated data."

There is an agreement among researchers about the following characteristics of VL:

- It is a kind of cognitive ability that also draws on the affective domain.
- Its focus is on intentional communication in an instructional context.

- It has accepted and incorporated theoretical contributions from other disciplines.
- It is described with terms, such as *ability*, *skill*, and *competency* (Avgerinou, 2009).

VL requires questioning, speculating, analyzing, fact-finding, and categorizing. By referring to competencies in using and interpreting still images, photography, film, video, and mass media, its early understanding often overlapped with media literacy and in many cases involved aesthetic appreciation, without discussing the creation or manipulation of images (Matusiak, Heinbach, Harper, & Bovee, 2019).

A cohesive VL theory should include visual perception, visual language, visual learning, visual thinking, and visual communication (Emanuel & Challons-Lipton, 2013). Visual literacy has been present since the 18th century, but gained momentum with the recent technological developments enabled by digital technologies that support access to large quantities of information and data (partially big data) and allows their easy manipulation, increasing in this way the interest in different kinds of data (Aldrich & Sheppard, 2000). Nevertheless, we should not forget that it is not limited to modern technologies.

Visual literacy skills are not isolated from other sensory skills (Avgerinou, 2009). These skills contribute to the development of critical thinking, deepening in this manner oral, reading, and writing literacy skills (Lopatovska, 2016).

Attention must be given to the need for learning how we process visual texts and meaning making is also a central object of VL that is often claimed to be a powerful aid to achieve it (Bawden & Robinson, 2016; Hobbs, 2006).

VL pertains to the class of literacies that concentrates on representation instead of emphasizing tool use. In this quality and by its subject matter, it is tied to media literacy. It is also closely related to information literacy. Its relationship to data literacy has special importance, because visualization is central to managing different kinds of data (Ridsdale et al., 2015) as data visualization allows us to discover the structure and patterns in datasets by mapping series of data to a space (Abilock et al., 2017). Therefore, visual literacy is a resource that can be integrated into data literacy. A survey on developing students' visual literacy skills by Schwartz (2018) revealed that VL is considered an important skill across disciplines. On the other hand, only 37.3% of the 118 respondents from academic libraries indicated that they or their institutions provide visual literacy education.

Graphic representations (visual depictions) of data and information include blueprints, caricatures, cartoons, charts, diagrams, drawings, flowcharts, graphs, maps, paintings, photographs, storyboard graphics, videos, mind maps, and concept maps (Metros, 2008). Visual techniques, such as mind mapping, can support cognitive processes such as thinking, problem solving, idea generation, categorization, representation, and vocabulary exploration (Fourie, 2011). Concept mapping can also be helpful, because concept maps foster individuals' ability to remember information, including its relationship to other concepts and facts (Schwendimann, 2015). Another and increasingly popular form of depiction is infographics (information graphics), produced with the intention to present information quickly and clearly. They present information and data in a concise and easily accessible manner. Unlike posters, which are usually purely informational as they present disjointed facts, infographics convey an overarching message (Weidler-Lewis, Lamb, & Polman, 2018).

Visual literacy requires the understanding of visual design, by taking account of the emotional, psychological, physiological, and cognitive influences in the perceptions of visual materials. It requires the comprehension of representational, explanatory, abstract, and symbolic images. It comprises data visualization, i.e., the process by which a visual image is formed. This process includes observation that is often followed by building abstract representations in a process, which facilitates reconceptualization and understanding of the previous experience before describing it verbally (Avgerinou, 2009).

According to Corrall (2017, p. 39), similarly to information literacy, visual literacy needs to change direction by giving less attention to "a functionalist, technological, skills-based and product-oriented approach." It should focus on a "situated, ecological, context- and process-oriented approach that embraces higher-order thinking and empowers students as knowledgeable critical information users."

Data visualization literacy

Data visualization literacy is tied to both data literacy and visual literacy. As Donohoe and Costello (2020) explain it, the need for and the practice of data visualization is growing. Varied forms of visualization represent data in an easy and understandable form. This is the reason why data visualization needs to be "suitable to extract the correct information to complete a task and make an informed decision while minimising the impact of biases" (p. 115). They also add that the perception of data visualization is affected by various factors, including education, politics, personal experience, usefulness, skills, and biases. Research shows that there are overconfident users of data visualization whose performance is impaired. On the other extreme, many persons are cautious about their visualization literacy skills, perceiving their own skills lower than they actually are.

Data visualization includes any kind of visual representation of digital data in computer-based systems. Even if infographics are not always tied to a computer-based projection, these can also be assigned to data visualization. Thus data visualization is a generic term for the visual projection of data on a two-dimensional plane, even if the representation simulates a three-dimensional space (Nazemi, Kaupp, Burkhard, & Below, 2021).

If examining data visualization literacy, we can also examine data-driven storytelling using visualization and infographics (Corrall, 2019). Taking into account that we as humans are open to sharing information in the form of stories, data-driven storytelling, also called data storytelling, is a methodology for communicating information tailored to a specific audience, with narratives (What is Data Storytelling?, n.d.). Data storytelling is related to data narratives that can be defined as a structured composition of messages that convey either findings over the data, and/or are delivered via visual means in order to facilitate their reception by an intended audience. When contemplating the levels of any narration, we ask two questions. When asking "what," we refer to some events that happen, and what is told is a story. The "how" question refers to the way and the form of telling, i.e., explaining how we presented it and transmitted it to our audience. The concept of discourse here answers to the "what" question. The story can be seen as the logical form of the narrative, while the discourse is its manifestation (El Outa, Francia, Marcel, Peralta, & Vassiliadis, 2020).

AI literacy

Two major goals of artificial intelligence (AI) are to:

- "Create expert systems, i.e., systems that exhibit intelligent behaviour, learn, demonstrate, explain, and advise their users.
- Implement human intelligence in machines, i.e., create systems that understand, think, learn, and behave like humans" (Kotis, 2021, p. 1).

The interest in AI (artificial intelligence) literacy is rooted in the theory and intense activity around big data. At present, AI is expensive and, more importantly, it should show improved transparency of systems and their design (Cox, Pinfield, & Rutter, 2019).

As a possible subprocess of approaching artificial intelligence, a critical approach to data can mobilize some kind of knowledge, similar to that originally devised as a component of media literacy and described by Potter (2018). This knowledge base consists of systematized sets of information that we build up and then store in our memory. It is made up of information elements put together according to a well-organized plan, and its structure helps us to recognize patterns.

Algorithms are described by many as black boxes, so they are considered opaque (Willson, 2017). This is because their operation is not transparent in many respects, it is difficult to use them to analyze phenomena, and drawing the boundaries of their decision-making mechanisms is a demanding task (Lloyd, 2017). We can agree therefore with the argument of Thomas, Nafus and Sherman (2018, p. 1), who say that algorithms are powerful "because we invest in them the power to do things. With such promise, they can transform the ordinary, say snapshots along a robotic vacuum cleaner's route, into something much more, such as a clean home."

Artificial intelligence (AI) is a highly disruptive technology, which can be understood as "giving machines the ability to interact with their environment and to make decisions: with a varying degree of autonomy, based on data collected or given to them, and in a manner which copies human thinking and can thus be considered as intelligent. Its impact on citizens, on companies, on public authorities and on society in general is the subject of much research but its impact on workers has been less of a focus" (Ponce del Castillo, 2020, p. 2).

AI literacy is a set of competencies that enables individuals to evaluate AI technologies critically. It allows communicating and collaborating effectively with AI and uses it as a tool online, at home, and in the workplace. As Long and Magerko (2020) stated, AI literacy is clearly related to a number of literacies, and AI-literate people need to understand how to use computers to make sense of AI. The other side of the coin is that data literacy is closely related to the AI subfield of machine learning, and therefore certain data literacy competencies overlap with AI literacy competencies. The potential prevalence of AI also may influence how we see information literacy, because it may mean "the ability to navigate a new space of AI tools and data, and data literacies, including critical awareness of how to protect one's own privacy" (Cox et al., 2019, p. 432).

In the AI environment, contemporary library roles of explaining how to navigate new information environments need to be extended to the understanding of the scholarly publishing landscape, including data creation processes (Cox et al., 2019). The efforts of the library community are fit for finding ways of incorporating data science into their internal operations and RDM services. Workforce development in libraries therefore may include not only data science, but also machine learning and artificial intelligence, supported by integrating algorithmic methods (Mani, Cawley, Henley, Triumph, & Williams, 2021).

Conclusion

We should listen to the advice of Bawden (2014), who suggested that we should not stop looking for new forms and concepts of literacies, because there is not only a proliferation of concepts in this area, but the contexts surrounding us are genuinely complex and consist of changing sets of issues. Convergences probably never end, as old convergences can be followed by new ones.

References

Abilock, D., Ballard, S. D., Bergson-Michelson, T., Colby, J., Fontichiaro, K., Hoelter, L., … Steelberg, T. (2017). *Creating data literate students*. Ann Arbor, MI: Michigan Publishing.

Abner, K. (2020). Data literacy as digital humanities literacy: Exploration of threshold concepts. In M. Brooks, M. Hubbard, J. Perkins, & J. Russell (Eds.), *Literacies in a digital humanities context: A dh + lib special issue* (pp. 19–22). dh + lib.

ACRL. (2000). *Information literacy competency standards for higher education*. Chicago: Information literacy: Association of College and Research Libraries. Retrieved from http://www.ala.org/acrl/standards/informationliteracycompetency.

ACRL. (2011). *ACRL visual literacy competency standards for higher education*. Chicago, IL: Association of College and Research Libraries. Retrieved from http://www.ala.org/acrl/standards/visualliteracy.

ACRL. (2013). *Intersections of scholarly communication and information literacy: Creating strategic collaborations for a changing academic environment*. Chicago, IL: Association of College and Research Libraries. http://acrl.ala.org/intersections.

ACRL. (2016). *Framework for information literacy for higher education*. Chicago, IL: Association of College and Research Libraries. Retrieved from http://www.ala.org/acrl/standards/ilframework.

Aldrich, F., & Sheppard, L. (2000). Graphicacy; the fourth 'R'? *Primary Science Review*, *64*, 8–11.

Andretta, S., Pope, A., & Walton, G. (2008). Information literacy education in the UK. *Communications in Information Literacy*, 2(1), 36–51. https://doi.org/10.15760/comminfolit.2008.2.1.55.

Avgerinou, M. D. (2009). Re-viewing visual literacy in the "Bain d'Images" era. *TechTrends*, *53*(2), 28–34. https://doi.org/10.1007/s11528-009-0264-z.

Bawden, D. (2014). Being fluent and keeping looking. In S. Kurbanoğlu, S. Špiranec, E. Grassian, D. Mizrachi, & R. Catts (Eds.), *Lifelong learning and digital citizenship in the 21st century* (pp. 13–18). Cham: Springer International Publishing.

Bawden, D., & Robinson, L. (2016). Information and the gaining of understanding. *Journal of Information Science*, *42*(3), 294–299. https://doi.org/10.1177/0165551515621691.

Bawden, D., & Robinson, L. (2017). 'An intensity around information': The changing face of chemical information literacy. *Journal of Information Science*, *43*(1), 17–24. https://doi.org/10.1177/0165551515616919.

Berry, D. (2011). The computational turn: Thinking about the digital humanities. *Culture Machine*, *12*. http://svr91.edns1.com/~culturem/index.php/cm/article/view/440/470.

Boyd, D., & Crawford, K. (2012). Critical questions for big data: Provocations for a cultural, technological, and scholarly phenomenon. *Information, Communication & Society*, *15*(5), 662–679. https://doi.org/10.1080/1369118X.2012.678878.

Budd, J. M. (2020). Information literacy and consciousness. *Journal of Documentation*, *76*(6), 1377–1391. https://doi.org/10.1108/JD-02-2020-0026.

Bundy, A. (2004). *Australian and New Zealand information literacy framework. Principles, standards and practice*. Adelaide: Australian and New Zealand Institute for Information Literacy.

Bundy, A. (2007). Computational thinking is pervasive. *Journal of Scientific and Practical Computing*, *1*(2), 67–69.

Calzada Prado, J. C., & Marzal, M.Á. (2013). Incorporating data literacy into information literacy programs: Core competencies and contents. *Libri*, *63*(2), 123–134. https://doi.org/10.1515/libri-2013-0010.

Carlson, J., Fosmire, M., Miller, C. C., & Sapp Nelson, M. (2011). Determining data information literacy needs: A study of students and research faculty. *Libraries and the Academy, 11*(2), 629–657. https://doi.org/10.1353/pla.2011.0022.

Carlson, J., & Johnston, L. R. (2015). *Data information literacy: Librarians, data, and the education of a new generation of researchers*. West Lafayette, IN: Purdue University Press.

Chauvin, B. A. (2003). Visual or media literacy? *Journal of Visual Literacy, 23*(2), 119–128. https://doi.org/10.1080/23796529.2003.11674596.

CILIP. (2018). *CILIP definition of information literacy 2018*. https://infolit.org.uk/information-literacydefinitionCILIP2018.pdf.

Corrall, S. (2017). Crossing the threshold: Reflective practice in information literacy development. *Journal of Information Literacy, 11*(1), 23–53. http://dx.doi.org/10.11645/11.1.2241.

Corrall, S. (2019). Repositioning data literacy as a mission-critical competence. In *ACRL 2019: Recasting the narrative*. http://d-scholarship.pitt.edu/id/eprint/36975.

Cox, A. (2018). Academic librarianship as a data profession: The familiar and unfamiliar in the data role spectrum. eLucidate, *15*(1/2). Retrieved from https://elucidate-ukeig.org.uk/index.php/elucidate/article/view/251.

Cox, A. M., Pinfield, S., & Rutter, S. (2019). The intelligent library. *Library Hi Tech, 37*(3), 418–435. https://doi.org/10.1108/LHT-08-2018-0105.

Davies, A., Fidler, D., & Gorbis, M. (2011). *Future work skills 2020*. Institute for the Future for University of Phoenix Research Institute. 540. Retrieved from https://www.voced.edu.au/content/ngv:49812.

Donohoe, D., & Costello, E. (2020). Data visualisation literacy in higher education: An exploratory study of understanding of a learning dashboard tool. *International Journal of Emerging Technologies in Learning, 15*(17), 115–126. https://doi.org/10.3991/ijet.v15i17.15041.

Dora, M., & Kumar, H. A. (2019). National and international trends in library and information science research: A comparative review of the literature. *IFLA Journal*. https://doi.org/10.1177/0340035219886610.

Dowell, M. M. S. (2019). Toward a working definition of digital literacy. In *Advanced methodologies and technologies in library science, information management, and scholarly inquiry* (pp. 118–129). Hershey, PA: IGI Global. https://doi.org/10.4018/978-1-5225-7659-4.ch01.

El Outa, F., Francia, M., Marcel, P., Peralta, V., & Vassiliadis, P. (2020). Towards a conceptual model for data narratives. In G. Dobbie, U. Frank, G. Kappel, S. W. Liddle, & H. C. Mayr (Eds.), *Conceptual modeling* (pp. 261–270). Cham: Springer.

Emanuel, R., & Challons-Lipton, S. (2013). Visual literacy and the digital native: Another look. *Journal of Visual Literacy, 32*(1), 7–26. https://doi.org/10.1080/23796529.2013.11674703.

Farrell, R., & Badke, W. (2015). Situating information literacy in the disciplines: A practical and systematic approach for academic librarians. *Reference Services Review, 43*(2), 319–340. https://doi.org/10.1108/RSR-11-2014-0052.

Federer, L. (2016). Research data management in the age of big data: Roles and opportunities for librarians. *Information Services & Use, 36*(1/2), 35–43.

Federer, L., Clarke, S. C., & Zaringhalam, M. (2020). *Developing the librarian workforce for data science and open science*. Bethesda, MD: National Library of Medicine. https://osf.io/uycax/.

Fister, B. (2017). The warp and weft of information literacy: Changing contexts, enduring challenges. *Journal of Information Literacy, 11*(1), 68–79. https://doi.org/10.11645/11.1.2183.

Fontichiaro, K., & Johnston, M. P. (2020). Rapid shifts in educators' perceptions of data literacy priorities. *Journal of Media Literacy Education, 12*(3), 75–87. https://doi.org/10.23860/JMLE-2020-12-3-7.

Fontichiaro, K., & Oehrli, J. A. (2016). Why data literacy matters. *Knowledge Quest, 44*(5), 22–27.

Fosmire, M., & Miller, C. (2008). Creating a culture of data integration and interoperability: Librarians and Earth Science Faculty collaborate on a geoinformatics course. In *Proceedings of the IATUL conferences*. Paper 16, Retrieved from http://docs.lib.purdue.edu/iatul/2008/papers/16.

Fourie, I. (2011). Personal information management (PIM), reference management and mind maps: The way to creative librarians? *Library Hi Tech*, *29*(4), 764–771. https://doi.org/10.1108/07378831111189822.

Gebre, E. H. (2018). Young adults' understanding and use of data: Insights for fostering secondary school students' data literacy. *Canadian Journal of Science, Mathematics and Technology Education*, *18*(4), 330–341. https://doi.org/10.1007/s42330-018-0034-z.

Gebre, E. H., & Morales, E. (2020). How 'accessible' is open data? Analysis of context-related information and users' comments in open datasets. *Information and Learning Sciences*, *121*(1/2), 19–36. https://doi.org/10.1108/ILS-08-2019-0086.

Gray, J., Gerlitz, C., & Bounegru, L. (2018). Data infrastructure literacy. *Big Data & Society*, *5*(2). https://doi.org/10.1177/2F2053951718786316.

Grillenberger, A., & Romeike, R. (2018). Developing a theoretically founded data literacy competency model. In *WiPSCE '18 proceedings of the 13th workshop in primary and secondary computing education* (pp. 1–10). https://doi.org/10.1145/3265757.3265766.

Haendel, M. A., Vasilevsky, N. A., & Wirz, J. A. (2012). Dealing with data: A case study on information and data management literacy. *PLoS Biology*, *10*, e1001339. https://doi.org/10.1371/journal.pbio.1001339.

Hawkley, E. (2017). Where's Walden? Searching, googling, reading, and living in the digital age. *Scholarly and Research Communication*, *8*(2). https://doi.org/10.22230/src.2017v8n2a277.

Hobbs, R. (2006). Multiple visions of multimedia literacy: Emerging areas of synthesis. In M. C. McKenna (Ed.), *Vol. 2. International handbook of literacy and technology* (pp. 15–28). Mahwah, NJ: Lawrence Erlbaum Associates. https://doi.org/10.1108/JD-02-2014-0026.

Hofer, A. R., Townsend, L., & Brunetti, K. (2012). Troublesome concepts and information literacy: Investigating threshold concepts for IL instruction. *Portal: Libraries and the Academy*, *12*(4), 387–405. https://doi.org/10.1353/pla.2012.0039.

Hunt, K. (2005). The challenges of integrating data literacy into the curriculum in an undergraduate institution. *IASSIST Quarterly*, *28*(2–3), 12–16.

Husfeldt, V., & Alt, S. (2021). From information literacy to data literacy education: A case study from Switzerland. In T. Schmidt, & C. Wolff (Eds.), *Information between data and Knowledge. Information science and its neighbors from data science to digital humanities (pp. 73–77) Proceedings of the 16th international symposium of information science (ISI 2021), Regensburg, Germany, 8–10 March*. Glückstadt: Verlag Werner Hülsbusch. https://doi.org/10.5283/epub.44938.

Ince, S., Hoadley, C., & Kirschner, P. (2018). The role of libraries in teaching doctoral students to become information-literate researchers: A review of existing practices and recommendations for the future. *Information and Learning Science*, *120*(3/4), 158–172. https://doi.org/10.1108/ILS-07-2018-0058.

Jacob, S. R., & Warschauer, M. (2018). Computational thinking and literacy. *Journal of Computer Science Integration*, *1*(1). https://doi.org/10.5283/10.26716/jcsi.2018.01.1.1.

Johnston, B., & Webber, S. (2003). Information literacy in higher education: A review and case study. *Studies in Higher Education*, *28*(3), 335–352. https://doi.org/10.1080/03075070309295.

Kitchin, R. (2014). Big data, new epistemologies and paradigm shifts. *Big Data & Society*, *1*(1). https://doi.org/10.1177/2053951714528481.

Koesten, L., Gregory, K., Groth, P., & Simperl, E. (2021). Talking datasets—Understanding data sensemaking behaviours. *International Journal of Human-Computer Studies*, *146*, 102562. https://doi.org/10.1016/j.ijhcs.2020.102562.

Koltay, T. (2015). Data literacy: In search of a name and identity. *Journal of Documentation, 71*(2), 401–415. https://doi.org/10.1108/JD-02-2014-0026.
Koltay, T. (2017). Data literacy for researchers and data librarians. *Journal of Librarianship and Information Science, 49*(1), 3–4. https://doi.org/10.1177/0961000615616450.
Koltay, T., Špiranec, S., & Karvalics, L. Z. (2016). *Research 2.0 and the future of information literacy*. Oxford: Chandos Publishing.
Kotis, K. (2021). Artificial general intelligence and creative economy. *Academia Letters*. https://doi.org/10.20935/AL260. Article 260.
Lau, J. (2013). Conceptual relationship of information literacy and media literacy. In A. Y. Lee, J. Lau, T. Carbo, & N. Gendina (Eds.), *Conceptual relationship of information literacy and media literacy in knowledge societies* (pp. 76–91). Paris: UNESCO.
Leaning, M. (2013). Towards the integration of media and information literacy: A rationale for a 21st century approach. In P. Mihailidis, & B. De Abreu (Eds.), *Media literacy education in action: Theoretical and pedagogical perspectives* (pp. 4–13). London: Routledge.
Lee, A. Y. (2013). Literacy and competencies required to participate in knowledge societies. In A. Y. Lee, J. Lau, T. Carbo, & N. Gendina (Eds.), *Conceptual relationship of information literacy and media literacy in knowledge societies* (pp. 3–75). Paris: UNESCO.
Lee, A. Y., & So, C. Y. (2014). Media literacy and information literacy: Similarities and differences. *Comunicar, 21*(42). https://doi.org/10.3916/C42-2014-13.
Leu, D. J., Kinzer, C. K., Coiro, J., Castek, J., & Henry, L. A. (2017). New literacies: A dual-level theory of the changing nature of literacy, instruction, and assessment. *Journal of Education, 197*(2), 1–18. https://doi.org/10.1177/002205741719700202.
Leu, D. J., Kinzer, C. K., Coiro, J. L., & Cammack, D. W. (2004). Toward a theory of new literacies emerging from the internet and other information and communication technologies. *Theoretical Models and Processes of Reading, 5*(1), 1570–1613.
Li, Y., Chen, Y., & Wang, Q. (2021). Evolution and diffusion of information literacy topics. *Scientometrics*, 1–30. https://doi.org/10.1007/s11192-021-03925-y.
Livingstone, S., van Couvering, E. J., & Thumin, N. (2008). Converging traditions of research on media and information literacies: Disciplinary and methodological issues. In J. Coiro, M. Knobel, C. Lankshear, & D. J. Leu (Eds.), *Handbook of research on new literacies* (pp. 13–18). Hillsdale, NJ: Lawrence Erlbaum Associates.
Llebot, C., & Rempel, H. G. (2021). Why won't they just adopt good research data management practices? An exploration of research teams and librarians' role in facilitating RDM adoption. *Journal of Librarianship and Scholarly Communication, 9*(General Issue), eP2321. https://doi.org/10.7710/2162-3309.2321.
Lloyd, A. (2017). Information literacy and literacies of information: A mid-range theory and model. *Journal of Information Literacy, 11*(1), 91–105. https://doi.org/10.11645/11.1.2185.
Long, D., & Magerko, B. (2020). What is AI literacy? Competencies and design considerations. In *The 38th annual CHI conference on human factors in computing systems. April 25–30, 2020, Honolulu, HI*. https://doi.org/10.1145/3313831.3376727.
Lopatovska, I. (2016). Engaging young children in visual literacy instruction. *Proceedings of the Association for Information Science and Technology, 53*(1), 1–5. https://doi.org/10.1002/pra2.2016.14505301101.
Macy, K. V., & Coates, H. L. (2016). Data information literacy instruction in business and public health: Comparative case studies. *IFLA Journal, 42*(4), 313–327. https://doi.org/10.1177/0340035216673382.
Mani, N. S., Cawley, M., Henley, A., Triumph, T., & Williams, J. M. (2021). Creating a data science framework: A model for academic research libraries. *Journal of Library Administration*. https://doi.org/10.1080/01930826.2021.1883366.
Mason, J., Khan, K., & Smith, S. (2016). Literate, numerate, discriminate: Realigning 21st century skills. In *Proceedings of the 24th international conference on computers in education* (pp. 609–614). Asia-Pacific Society for Computers in Education.

Matusiak, K. K., Heinbach, C., Harper, A., & Bovee, M. (2019). Visual literacy in practice: Use of images in students' academic work. *College, & Research Libraries*, *80*(1), 123–139. https://doi.org/10.5860/crl.80.1.123.

Maybee, C., & Zilinski, L. (2015). Data informed learning: A next phase data literacy framework for higher education. In *Proceedings of the 78th ASIS&T annual meeting: Information science with impact: Research in and for the community* (pp. 108–111). Silver Spring, MD: American Society for Information Science.

Metros, S. E. (2008). The educator's role in preparing visually literate learners. *Theory Into Practice*, *47*(2), 102–109. https://doi.org/10.1080/00405840801992264.

Meyer, J., Land, R., & Baillie, C. (2010). *Threshold concepts and transformational learning*. Rotterdam, Boston: Sense Publishers.

Nazemi, K., Kaupp, L., Burkhard, D., & Below, N. (2021). Datenvisualisierung. In M. Putnings, H. Neuroth, & J. Neumann (Eds.), *Praxishandbuch Forschungsdatenmanagement* (pp. 477–502). Berlin: De Gruyter Saur. https://doi.org/10.1515/9783110657807.

Nicholas, D., et al. (2008). The information-seeking behaviour of the digital consumer: Case study the virtual scholar. In D. Nicholas, & I. Rowlands (Eds.), *Digital consumers: Reshaping the information professions* (pp. 113–158). London: Facet.

Norgaard, R., & Sinkinson, C. (2016). Writing information literacy: A retrospective and a look ahead. In *Information literacy: Research and collaboration across disciplines* (pp. 15–35). Fort Collins, CO: WAC Clearinghouse and University Press of Colorado.

Ojanen, M., Lindholm, T., & Siipilehto, L. (2020). *What is research data management (RDM)?* Retrieved from https://blogs.helsinki.fi/thinkopen/know-your-data-rdm-series-1/.

Oliphant, T. (2017). A case for critical data studies in library and information studies. *Journal of Critical Library and Information Studies*, *1*(1). https://doi.org/10.24242/jclis.v1i1.22.

Onyancha, O. B. (2020). Knowledge visualization and mapping of information literacy, 1975–2018. *IFLA Journal*, *46*(2), 107–123. https://doi.org/10.1177/0340035220906536.

Pangrazio, L., & Sefton-Green, J. (2020). The social utility of 'data literacy'. *Learning, Media and Technology*, *45*(2), 208–220. https://doi.org/10.1080/17439884.2020.1707223.

Pinto, M., Pulgarín, A., & Escalona, M. I. (2014). Viewing information literacy concepts: A comparison of two branches of knowledge. *Scientometrics*, *98*(3), 2311–2329. https://doi.org/10.1007/s11192-013-1166-6.

Ponce del Castillo, A. (2020). *Labour in the age of AI: Why regulation is needed to protect workers*. Brussels: ETUI. ETUI Research Paper-Foresight Brief 08. Retrieved from https://www.etui.org/publications/foresight-briefs/labour-in-the-age-of-ai-why-regulation-is-needed-to-protect-workers.

Potter, J. W. (2018). *Media literacy*. CA, Sage: Thousand Oaks.

Potter, J. W. (2019). *The seven skills of media literacy*. CA, Sage: Thousand Oaks.

Qin, J., & D'Ignazio, J. (2010). Lessons learned from a two-year experience in science data literacy education. In *Proceedings of the 31st annual IATUL conference, June, 20-24, 2010*. Retrieved from http://docs.lib.purdue.edu/iatul2010/conf/day2/5.

Ramsden, D. (2013). *The age of convergence*. New York, NY: DreamTiger Equities.

Ridsdale, C., Rothwell, J., Smit, M., Ali-Hassan, H., Bliemel, M., & Irvine, D. (2015). *Strategies and best practices for data literacy education. Knowledge synthesis report*. Halifax, NS: Dalhousie University.

RIN. (2011). *The role of research supervisors in information literacy*. Research Information Network. Retrieved from http://www.rin.ac.uk/system/files/attachments/Research_supervisors_report_for_screen.pdf.

Robinson, L. (2009). Information science: Communication chain and domain analysis. *Journal of Documentation*, *65*(4), 578–591. https://doi.org/10.1108/00220410910970267.

Sander, I. (2020). What is critical big data literacy and how can it be implemented? *Internet Policy Review*, *9*(2). https://doi.org/10.14763/2020.2.1479.

Schield, M. (2004). Information literacy, statistical literacy and data literacy. *IASSIST Quarterly*, *28*(2/3), 6–11.

Schield, M. (2010). Assessing statistical literacy: Take CARE. In P. Bidgood, N. Hunt, & F. Jolliffe (Eds.), *Assessment methods in statistical education: An international perspective* (pp. 133–152). New York, NY: Wiley.

Schneider, R. (2013). Research data literacy. In *Worldwide commonalities and challenges in information literacy research and practice* (pp. 134–140). Cham: Springer International.

Schüller, K. (2020). *Future skills: A framework for data literacy*. Working Paper No. 53 Berlin: Hochschulforum Digitalisierung. https://doi.org/10.5281/zenodo.3946067.

Schüller, K., & Busch, P. (2019). *Data literacy: Ein systematic review zu begriffsdefinition, kompetenzrahmen und testinstrumenten*. Arbeitspapier Nr. 46 Berlin: Hochschulforum Digitalisierung. https://doi.org/10.5281/zenodo.3349865.

Schüller, K., Busch, P., & Hindinger, C. (2019). *Future skills. Ein framework für data literacy. Kompetenzrahmen und Forschungsbericht*. Forschungsbericht Nr. 47 Berlin: Hochschulforum Digitalisierung. https://doi.org/10.5281/zenodo.3349865 (in German).

Schwartz, J. (2018). Visual literacy: Academic libraries address 21st century challenges. *Reference Services Review*, *46*(4), 479–499.

Schwendimann, B. A. (2015). Concept maps as versatile tools to integrate complex ideas: From kindergarten to higher and professional education. *Knowledge Management and E-Learning*, *7*(1), 73–99. https://doi.org/10.34105/j.kmel.2015.07.006.

SCONUL. (2011). *The SCONUL seven pillars of information literacy. Core model for higher education*. London: Society of College, National and University Libraries Working Group on Information Literacy. Retrieved from http://www.sconul.ac.uk/sites/default/files/documents/coremodel.pdf.

Shome, A. (2019). Closing the gap: A practical guide to science in the writing center. *Praxis: A Writing Center Journal*, *17*(1). Retrieved from http://www.praxisuwc.com/171-shome.

Si, L., Zhuang, X., Xing, W., & Guo, W. (2013). The cultivation of scientific data specialists: Development of LIS education oriented to e-science service requirements. *Library Hi Tech*, *31*(4), 700–724. https://doi.org/10.1108/LHT-06-2013-0070.

Song, I. Y., & Zhu, Y. (2017). Big data and data science: Opportunities and challenges of iSchools. *Journal of Data and Information Science*, *2*(3), 1–18. https://doi.org/10.1515/jdis-2017-0011.

Špiranec, S., Banek Zorica, M., & Kos, D. (2016). Information literacy in participatory environments: The turn towards a critical literacy perspective. *Journal of Documentation*, *72*(2), 247–264. https://doi.org/10.1108/JD-06-2015-0072.

Špiranec, S., Kos, D., & George, M. (2019). Searching for critical dimensions in data literacy. *Information Research*, *24*(4). paper colis 1922 http://InformationR.net/ir/24-4/colis/colis1922.html.

Stephenson, E., & Schifter Caravello, P. (2007). Incorporating data literacy into undergraduate information literacy programs in the social sciences: A pilot project. *Reference Services Review*, *35*(4), 525–540. https://doi.org/10.1108/00907320710838354.

Tenopir, C., Sandusky, R. J., Allard, S., & Birch, R. (2014). Research data management services in academic research libraries and perceptions of librarians. *Library & Information Science Research*, *36*(3), 84–90. https://doi.org/10.1016/j.lisr.2013.11.003.

Tewell, E. (2015). A decade of critical information literacy: A review of the literature. *Communications in Information Literacy*, *9*(1), 2. https://doi.org/10.15760/comminfolit.2015.9.1.174.

Thomas, S. L., Nafus, D., & Sherman, J. (2018). Algorithms as fetish: Faith and possibility in algorithmic work. *Big Data & Society*, *5*(1). https://doi.org/10.1177/2053951717751552.

Thompson, K. A., & Edelstein, D. M. (2004). A reference model for providing statistical consulting services in an academic library setting. *IASSIST Quarterly*, *28*(2), 35–38.

Thoreau, H. D. (1995). *Walden; or, life in the woods*. New York, NY: Dover Publications.

Townsend, L., Brunetti, K., & Hofer, A. R. (2011). Threshold concepts and information literacy. *Portal: Libraries and the Academy*, *11*(3), 853–869. https://doi.org/10.1353/pla.2011.0030.

Vitae Researcher Development Framework. (2012). https://www.vitae.ac.uk/researchers-professional-development/about-the-vitae-researcher-development-framework/developing-the-vitae-researcher-development-framework.

Walton, G., & Cleland, J. (2017). Information literacy-empowerment or reproduction in practice? A discourse analysis approach. *Journal of Documentation*, *73*(4), 582–594. https://doi.org/10.1108/JD-04-2015-0048.

Walton, G., Pickard, A. J., & Dodd, L. (2018). Information discernment, mis-information and pro-active scepticism. *Journal of Librarianship and Information Science*, *50*(3), 296–309. https://doi.org/10.1177/0961000618769980.

Weidler-Lewis, J., Lamb, G. R., & Polman, J. (2018). Creative visual representation: Using science infographics to jump-start creativity in the classroom. *The Science Teacher*, *86*(2), 41–47.

What is Data Storytelling? (n.d.). Retrieved from https://www.nugit.co/what-is-data-storytelling/.

Willson, M. (2017). Algorithms (and the) everyday. *Information, Communication & Society*, *20*(1), 137–150.

Wilson, C. (2012). Media and information literacy: Pedagogy and possibilities. *Comunicar*, *20*(2), 15–22. https://doi.org/10.3916/C39-2012-02-01.

Wing, J. M. (2006). Computational thinking. *Communications of the ACM*, *49*(3), 33–35.

Wing, J. M. (2008). Computational thinking and thinking about computing. *Philosophical Transactions of the Royal Society of London A: Mathematical, Physical and Engineering Sciences*, *366*(1881), 3717–3725. https://doi.org/10.1098/rsta.2008.0118.

Zhang, L., & Eichmann-Kalwara, N. (2019). Mapping the scholarly literature found in Scopus on "research data management": A bibliometric and data visualization approach. *Journal of Librarianship and Scholarly Communication*, *7*(1), eP2226. https://doi.org/10.7710/2162-3309.2226.

Further reading

Daraio, C., Lenzerini, M., Leporelli, C., Naggar, P., Bonaccorsi, A., & Bartolucci, A. (2016). The advantages of an ontology-based data management approach: Openness, interoperability and data quality. *Scientometrics*, *108*(1), 441–455. https://doi.org/10.1007/s11192-016-1913-6.

McCosker, A. (2017). Data literacies for the postdemographic social media self. *First Monday*, *22*(10). https://doi.org/10.5210/fm.v22i110.7307.

Pangrazio, L., & Selwyn, N. (2019). 'Personal data literacies': A critical literacies approach to enhancing understandings of personal digital data. *New Media & Society*, *21*(2), 419–437. https://doi.org/10.1177/1461444818799523.

Wibeck, V., & Linnér, B. O. (2021). Sense-making analysis: A framework for multi-strategy and cross-country research. *International Journal of Qualitative Methods*, *20*. https://doi.org/10.1177/1609406921998907. 1609406921998907.

CHAPTER

6

Roles and education of information and data professionals

Information and data professionals in academic libraries and beyond

The prevalence of research data and the appearance of new, related professional groups did not subvert academic librarianship. These evolving groups include data librarians and data scientists, but there are data curators and data stewards as well. These developments deeply influence the vision and processes of academic libraries, even though the changes induced by them appear at different speeds in different countries. There is also a difference in the extent that these tasks will become everyday occupations, especially because acceptance may depend on the perceptions of librarians, researchers, and teaching staff members (Cox, 2018). The changes in perception and behavior of data professionals obviously appear on different levels of detail and granularity. Nonetheless, these professionals are increasingly trained at the intersection of library and information science (LIS) and data science (DS), although we cannot say that there will be a fusion of these two. Suspecting that the potential readers of this book are much more familiar with the knowledge, skills, and duties of data librarians, and may have even gathered experience within this field, we will pay more attention to issues related to data science.

Data librarians

As Rice and Southall (2016, p. 17) put it, "data librarianship is concerned with the representation, organization, and dissemination of data, and the use of technologies to design research data management and

Research Data Management and Data Literacies
https://doi.org/10.1016/B978-0-12-824475-3.00006-0

data services" that may require detailed specialist knowledge about collections or giving advice on copyright.

Kennan (2017) underlines how important it is to understand the significance of discipline-specific research life cycles and cultures/processes. She also directs our attention to research method ethics, as well as intellectual property and licensing laws and access norms. While related work roles and job titles of data librarians may vary across different organizations, the need for understanding data is not the only requirement. This understanding rather should be extended to familiarity with the research environment of the given institution, related to policies of funding agencies as well as research evaluation activities.

Robinson and Bawden (2017, pp. 313–314) affirmed that information professionals should be aware of data-related functions in academic libraries independently of their position and scope of duties, or whether they are directly involved in these activities. They stated that "information professionals may need to become involved in data support, research data management, data curation, data governance, data quality evaluation, data citation, data literacy training and similar activities." Data professionals may be involved in devising research data policies, developing and managing data repositories, and data curation programs. Furthermore, teaching data literacy and associated skills have become a default occupation for them. Learning about the nature of data management plans (DMPs) and advising researchers on preparing them may be based on supporting materials, such as the *Checklist for a Data Management Plan*, provided by the Digital Curation Centre (DCC, 2013).

Librarians need to have a conceptual understanding of data, and be knowledgeable of the surrounding issues, as well as being able to find, extract, collect, clean, organize, analyze, and present data. The requirements include being familiar with the related legal requirements and ethical considerations. There is also a need of being able to get the data in the needed form, and to draw the meaningful and useful information from it.

A content analysis of job advertisements for data librarians in the US academic libraries by Khan and Du (2018, p. 6) identified the following percentages of knowledge and skills required of data librarians:

- Research assistance (100%),
- Critical thinking/problem solving (89%),
- Outreach (61%),
- Visualization (25%),
- Quantitative and qualitative data analysis (20%),
- Teaching (14%),
- Standards-based metadata schemata (8%),
- Data literacy (6%).

The issues related to proprietary or open source statistical software packages appeared in 70% of the ads, which shows their importance. Nonetheless, we will avoid giving details about them, because the offering of such software may be prone to rapid changes in the marketplace, not to mention their potential obsolescence.

In an analysis similar to the previous one, Eclevia, Fredeluces, Maestro, and Eclevia Jr. (2019) found a number of frequently required skills, of which communication and data management were the most relevant. Similarly, data management and data management planning and (the use of) data visualization tools were the most requested data librarian qualifications. The latter issue, i.e., data visualization, is a prominent topic covered throughout this book, not only regarding LIS programs (e.g., Khan, 2020), but in the case of data science (e.g., Semeler & Pinto, 2020) and the combination of DS and open science (e.g., Federer, Clarke, Zaringhalam, & Huerta, 2020). Some details about this appear in the subsection addressing education for data literacy. Despite the apparently high interest in statistical software packages, we will not provide details about them, for the same reasons described earlier.

The need for employing data librarians is clearly visible from the study of Gowen and Meier (2020), who tracked the changes in RDM services and staffing among the Association of American Universities (AAU) member libraries. Their results show that 34 posts of the 60 created in 2014 were retained in 2019. Moreover, 12 institutions added a data librarian between 2014 and 2019.

Data science and data scientists

In their well-known paper, Davenport and Patil (2012, p. 73) declared not only that being a data scientist is the sexiest job of the 21st century, but underlined that it is about "bringing structure to large quantities of formless data and making analysis possible," indicating therefore that it is more than formal data analysis. As in the case of DS, we are speaking about a relatively young discipline and it is difficult to paint a correct picture of those who cultivate it. It is even more difficult to separate promises about its future from reality.

What is certain that there is no reliable evidence that a unified definition of a data scientist exists. This conclusion was drawn by Baškarada and Koronios (2017, pp. 65–69). As they say, it is unrealistic to expect a single person to have the "same level of expertise in a number of distinct disciplines as specialized experts can," because data scientists work instead effectively in multidisciplinary teams. They name instead a host of data science roles. There are thus data engineers, whose role is much more important in a data science environment than in the context of business intelligence.

The core of the data science team is the statistician, who enables working as a team by building bridges between domain experts, data engineers, and computer scientists. Statisticians are required to refine and formalize questions and ideas from domain experts, request relevant data from data engineers, and guide computer scientists in relation to data analysis.

Computer scientists need to be competent in using software tools and technologies, as well as the relevant programming languages. Communicators should not be forgotten, because they are responsible for managing organizational change by effectively communicating with relevant decision makers, among others, about problems and opportunities as well as results. It is rather obvious that members of the team have a leader, who has some understanding of all the roles of the given team members, in order to manage resources, tasks, and deliverables.

A different taxonomy of data scientists was given by Rawlings-Goss (2019, pp. 7–9). As she explained it, in most cases the collective name "data scientists" is used for generalists, who are broadly trained, quick problem solvers, and equipped to work on every stage of the analytics lifecycle (including data acquisition, transformation, and analytics). Data generalists are the ones who need to be prepared to do anything. They are expected to have strong and broad technical skills, including mathematical techniques, comprehension of statistics, and technology-based skills. Data generalists need to have nontechnical, interpersonal skills, such as communication abilities, and strong teamwork capabilities. As we will see, these skills are decisive; thus we will discuss them in more detail in the subsection on data librarianship.

Data analysts' work involves translating statistical information into a language that everyday people can understand and use it as a basis for making practical decisions. In a way, similar to building architects, data architects design the structure needed to house data and ensure that it is effectively gathered, stored, and managed. Data engineers test, evaluate, and improve the solutions designed by data architects. They are required to have a "precise knowledge and understanding of software engineering and testing patterns, combined with coding experience." The data engineer's job is related to the maintenance and further development of the infrastructure that was designed by data architects.

The relationship between DS and data librarianship is two-faced. Many of the related tasks suggest that connections to RDM may exist, even though the nature of the relationship between them is not always clear-cut. However, the philosophies behind them are different.

For instance, by underlining the importance of working with algorithms, machine learning, and statistical techniques, data science can be seen as "the blending of competencies in computer programming and software engineering, statistics, and a particular domain expertise" (Burton, Lyon, Erdmann, & Tijerina, 2018, p. 7). On the other hand, data scientists should be made responsible for the integrity of the research process, for which they must learn the vocabulary and the jargon of the given discipline and acquire the difficult skill of asking the right questions (Scroggins et al., 2019). They also need to understand the roles of big data and related technologies, and "work with all the steps of a data science lifecycle—discover and solve problems, and communicate solutions and use a range of tools to solve big data problems" (Song & Zhu, 2017, p. 5). These functions are clearly tied to RDM, although their relationship often remains opaque in many aspects.

A seemingly distant problem, related to bad big data in RDM systems, is also tied to data quality, addressed in Chapter 3 of this book, and illustrates this uncertainty. We know that results should be based only on flawless research. However, while research is being organized in self-contained projects, digital sources might enable each individual project to be part of a continuously progressing process of knowledge (Azeroual, 2020). However, the increasing amount of data can lead to more quality errors, both in the cases where data is recorded manually and with automated data collection processes (Azeroual & Lewoniewski, 2020).

The question about the scope of data scientists' expertise remains open, as it combines numerous skills and various fields of knowledge. This expertise allows "for the thinking or prediction of prototypes and demonstrations that help to illustrate data-intensive research." We know obviously that the "everyday" duties of data scientists may help create insights by triggering "new analytical activities and may completely change how things are done in the future." It is beneficial for data scientists to have experience in dealing with open-ended questions and "the ability to quickly iterate over different types of analysis methods and models." DS also requires familiarity with systems design, distributed computing, industrial engineering, and big data (Semeler & Pinto, 2020, p. 4).

Franklin and Nicolae (2021, p. 3) direct our attention to the potentially negative effects of data-driven systems. This implies that we should be aware that "in addition to computation and inference, the human context of data-driven decision making must be a key component of data science." No matter "whether such decisions are made by humans or by algorithms, it is humans and society that will ultimately have to deal with their consequences."

It is almost self-explanatory that—in general—data scientists are required to learn from experiences, as well as apply "out-of-the-box thinking and the ability to move beyond an existing blueprint" (Berthold, 2019, pp. 4–5).

An analysis of papers published in the Web of Science database from the LIS perspective showed that DS is addressed, among others, as a discipline (set of disciplines), a process (set of processes), a set of principles, and (obviously) a research field. This confirms what we said earlier, namely that DS is interdisciplinary by nature, i.e., involves knowledge and skills of LIS, business and economics, and computer science, just to name a few. It also turned out that DS can be distilled to accomplishing the following, partially overlapping, goals and processes:

- Collecting and analyzing data within a problem domain,
- Collecting, preparing, analyzing, visualizing, managing, and preserving large collections of information,
- Discovering meaningful patterns in data by using methods of computational analytics,
- Extracting nonobvious and useful patterns from large data,
- Providing techniques, methods, and tools to gain insights from new and existing datasets,
- Using processes and systems for extracting information, data, and knowledge from massive amounts of data,
- Solving big data challenges,
- Supporting and guiding the extraction of information and knowledge from data (Virkus & Garoufallou, 2019).

In our opinion, achieving some of these tasks is attainable within the framework of RDM. This is especially true for collecting and analyzing data, although the extent of this activity may be less in the case of RDM than in using DS methods.

Using the same database, Virkus and Garoufallou (2020) have found that from 2350 publications that appeared between 1980 and 2019, the contribution of LIS to data science was limited to 3.4%.

Working with research data requires adopting practices that encourage reproducibility (e.g., well-documented data collection, following data management best practices), which is also an enabler of good DS that is necessary for verifying whether DS tools are applicable to the data and the problem at hand (Oliver, Kollen, Hickson, & Rios, 2019). On the other hand, the tantalizing question about the similarity or convergence between the traditional, though changing profession of the data librarian and the evolving occupations appear from time to time. Both professions are being shaped by conditions and requirements of the data-intensive era. Furthermore, there are undeniable signs of similarities between these groups' profile and education. Nonetheless, we see that DS professionals and data librarians still have different scope of duties, work environment, and culture. As stated by Federer, Clarke, Zaringhalam, and Huerta (2020, p. 9) "definitions of data science and the specifics of its application to scientific problems may differ from one discipline to another, or even one research team to another."

Coupling DS and open science is possible due to the existence of knowledge and skills shared among a number of data librarians and data scientists. Such elements include hacking skills, quantitative analysis, and the abilities for extracting and structuring data (Semeler & Pinto, 2020).

Moreover, the relationship between data librarians and data scientists is in many regards unclear. Be that as it may, the most realistic general view on this issue is that we can still count on competition between them, but a wide range of data-related issues may require cooperation between them.

DS may provide new methods and practices for data librarianship, without requiring data librarians to become programmers, statisticians, or database managers (Koltay, 2019). We also may accept that DS "typically involves computational approaches to research that enable discovering meaning" (Federer, Clarke, Zaringhalam, & Huerta (2020, p. 9). Somewhat similarly, Franklin and Nicolae (2021, p. 4) state that the goal of DS is "to inform and influence decisions, then a deep understanding of how to present and consume data-driven arguments is necessary." These authors add that this is "not simply about the informing power of data, but it is a kind of data literacy."

One of the interfaces in the education for these two disciplines seems to be appearing through the iSchool movement (https://ischools.org/The-iSchool-Movement), because a number of higher education institutions originally offering LIS degrees have realized that their programs were capable of reaching a broader audience of students and to prepare information professionals for work beyond libraries.

The visions and ways of teaching DS in iSchools vary considerably. For instance, Song and Zhu (2017, p. 6) underline the importance of being able "to solve problems and create values, information, and knowledge using tools in application domains," emphasizing users, tools, and applications. Due to the complexity and highly interdisciplinary nature of DS, students should learn how to work with domain experts and students of other disciplines as a team.

Emphasizing that LIS can make unique contributions to data science research, Wang (2018) argued that iSchools should integrate both data science and LIS into their curricula. Virkus and Garoufallou (2019) also recognized the potential of making use of reciprocity in the sense that LIS educators and practitioners could understand the challenges and opportunities related to data science, while LIS can contribute to data science in the field of data curation and management, data analytics, visualization, and presentation.

As mentioned earlier, interpersonal (soft, nontechnical) skills are indispensable both for data scientists and data librarians, as explained by Kennan (2016), who mentions not only the skills related to advocacy, negotiation, and capability-building skills, but directs our attention to the need of being comfortable with change and having a service philosophy, willingness to learn, discretion, and curiosity. She also identified the need for being assertive and open to new experiences, having networking abilities, being confident enough to see the limits of one's own knowledge, and being able to rely on people with different knowledge or skills in the right moment.

The survey of librarians fulfilling data-related duties revealed that communication skills, the ability to develop relationships with researchers, and other interpersonal skills are especially relevant to their work (Federer, 2018).

As Henderson (2020, pp. 2–4) put it, "If you want to work with data, you will need to be an administrator and an entrepreneur, as well as a reference and instruction librarian, so data science knowledge is only a part of the job." She states accordingly that, in addition to technical skills such as programming or statistics, good listening without making assumptions is the foundation for building relationships, and fostering collaborations is indispensable. Furthermore, creating relationship maps is an excellent way to show progress when assessing RDS, because these maps can quantitatively express the strength of a relationship.

Federer, Clarke, Zaringhalam, and Huerta (2020, pp. 12–16) identified core skills that are needed for this reciprocity, adding that none of the library professionals would need to acquire all the following skills in order to be successful. Although neither is it likely that institutions would require their librarians to have all of them, the following list shows the complexity of these skills and abilities.

Data skills

- Data management plans and data workflows,
- Data and metadata standards and curation,
- Data sharing and reuse,
- Data citation,
- Data policy and governance,
- Computational skills,
- Computational literacy,
- Database design,
- Familiarity with relevant coding languages,
- Machine learning and data or text mining,

Computational skills

- Computational literacy,
- Database design,
- Machine learning and data or text mining,
- Data visualization (also pertaining to data skills),

Research and subject matter knowledge

- General understanding of the relevant science or subject matter,
- Research design and workflows,

- Statistics,
- Responsible conduct of research and research ethics,
- Methods for reproducibility,
- Traditional library skills,
- Consultation and reference,
- Literature searching,
- Scholarly communication,
- Bibliometrics,
- Training and instruction,
- Assessment and evaluation,

Skills for developing programs and services

- Interview and assessment skills to understand institutional needs,
- Scoping and planning for sustainability,
- Willingness to embrace failure,
- Communication and marketing skills,
- Interpersonal skills,
- Team science skills,
- Entrepreneurship,
- Advocacy skills,
- Community building,
- Skills for lifelong learning,
- Flexibility and adaptability,
- "Anthropological" mindset,
- Logic and problem-solving,
- Design thinking,
- Computational thinking.

The nature of these skills does not require detailed explanation. However, the concept of an "anthropological" mindset may sound unfamiliar, but it is about observing and adapting to the cultures and norms of different groups with whom one works.

A relatively early warning of Robinson (2016) related to the listed skills directs our attention to a different set of problems. She reminds us that being specialized in data, i.e., examining pattern and syntax and using quantitative methodology versus focusing on information, i.e., being interested in meaning, semantics, and using qualitative methods, results in becoming entrenched in professions that are defined in an overly narrow way, thus leading to silos. This division is harmful for both data scientists and data librarians. We will return to the problem of silos in a couple of cases in a small subsection of this chapter.

While stating that DS and LIS are twin disciplines by nature, Wang (2018) mentions that DS often ignores data quality, thus—if this situation continues—the old saying "garbage in, garbage out" from the 1980s will reappear in the big data age.

In order to identify DS courses in US LIS programs, Khan (2020) analyzed 128 syllabi. The most frequent course was data mining (96%), followed by analytical tools (93%), geographic information systems (78%), and data visualization (76%). Metadata was covered by 47% and data preservation (including storing and archiving) was present in 24% of the syllabi. Social, ethical, and legal issues were present with 88% in this investigation. This indicates again that we should move beyond technical skills.

The Data Science Association's Code of Professional Conduct follows this path by declaring that the role of the data scientist is giving meaning to raw data, because "the client is dependent on the manipulation and interpretation done by the scientist" (DSA, 2013). Although we know that the existence of raw data is not universally accepted, because data "can be good or bad, better or worse, incomplete and insufficient, but does not equal to facts (Gitelman, 2013, p. 7)."

More importantly, however, there are deeper key differences in approaching information and data. As stated by Trepanier, Shiri, and Samek (2019, pp. 292–294), "librarians and other information workers provide access to final form information, including helping to find/source it, but the user is responsible for interpreting the meaning that is attributable to it. The data scientist provides a 'value-added' service to data, librarians and other information workers connect the user to data/information, but do not enhance/interpret it in any way – in fact, are expected to refrain from judgement." They also direct our attention to another telling difference between the two approaches. Librarians work with "information that is publicly available (or within the realm of the institute under consideration), and therefore do not have a confidentiality issue with the information itself (although there is a responsibility not to identify the user with the information)." In contrast to this, data scientists work in explicit or implicit contractual relationships, which may specify or impose a level of exclusivity on the provision of data. As a rule, librarians act "as a public resource connecting all users, on an equitable basis, to any information that is available," and are not subject to concerns about conflict of interest. In the data science world, data "may be owned and capitalized upon for profit-maximization, and the ability to manipulate data can be a strategic asset to be offered to the marketplace, while librarians advocate for universal access whenever possible."

As a conclusion to these reflections, it seems to be useful to examine a Venn diagram, depicted in Fig. 1. This diagram was devised by Semeler and Pinto (2020), and describes the union and intersection between (library and) information science, e-science, and data science, representing in this way the theoretical and practical links between areas that make up data librarianship as an interdisciplinary field.

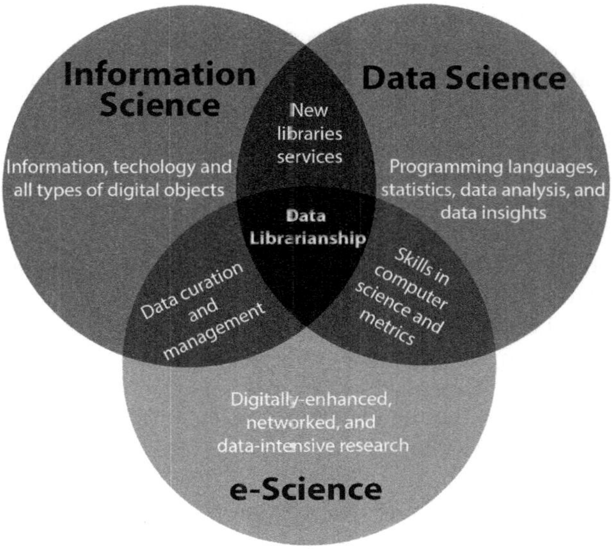

FIG. 1 Data librarianship Venn diagram. Adopted from Semeler, A. R., & Pinto, A. L. (2020). Data librarianship as a field. *Transinformação*, 32, e200034, https://doi.org/10.1590/2318-0889202032e200034, p. 3.

Other data professionals and their roles

Beyond these two groups of data professionals, we should include the work of *data curators*, who must be familiar with quality assurance standards, especially regarding authenticity, integrity, and accuracy of the data. They also look for "ways of accessing digital objects' authenticity, integrity, and accuracy over time, as well as storage and preservation policies, procedures and practices." These tasks pertain to the group of operational competencies, while the ability to understand and be able to communicate the risk of information loss or corruption of digital entities is a managerial competency (Madrid, 2013, p. 155).

Data curators' duties must be contrasted with curating collections in museums, where pieces are exhibited rather than kept for internal storage, whereas data needs to be preserved by frequent validation checks and backups. When defining digital curation and data curation, LIS education makes only minor distinctions between them by referring to "organizing, maintaining, integrating, and extracting knowledge from different types of data to apply and reuse it" (Thomas & Urban, 2018, p. 402).

The digital humanities represent an emerging conceptual and practical framework for digital curation in which theory, practice, collaboration, and digital content contribute to a rich foundation for new knowledge.

In the transdisciplinary digital environment, digital curation focuses on extending the life cycle of digital humanistic data (Sabharwal, 2017). Poole (2017) also emphasizes that digital curation and the digital humanities share concerns, practices, and objectives, among others, because both "add value to digital assets: they stimulate reuse, encourage the development of new research questions, and attract new audiences" (p. 1772). Apparently, there is again an intersection of academic librarianship, this time with DH.

Some of the duties of data curators and librarians can be related to working in data repositories, where they serve as primary expert contacts for new users when inquiring about submitting content to the data repository, and authorizing new submissions (Johnston et al., 2017).

Data stewards are designated subject matter experts, who have detailed knowledge about the data requirements and can translate these requirements into technical specifications (Abraham, Schneider, & Vom Brocke, 2019).

The need for good stewardship appeared when datasets grew larger and distributed collaborations became more common. The essence of stewardship is "releasing some kinds of data and preventing the release of other kinds of data. The same data may fall into either category, depending on the time, purpose, or entity requesting access" (Borgman, 2018, p. 368).

There is obviously a need for exercising some gatekeeping in order to ensure that data is easily accessible to those who need it and to keep it safe from unauthorized access, indicating an apparent intersection between the skills of data librarians and the work of data scientists. This convergence, already mentioned in this chapter, can positively influence the data librarian's work (Ogier & Stamper, 2018).

From among their varied roles, data stewards play an analytical role, which requires them to focus on data quality by ensuring informed decisions based on quality data. They are usually also responsible for making data compliant with the GDPR (General Data Protection Regulation) rules (Wildgaard, Vlachos, Nondal, Larsen, & Svendsen, 2020).

A Delphi study, involving 24 participants from different stakeholder groups in Norway, demonstrated that institutional closeness of data stewards is of prime importance, because their work requires detailed knowledge of research activities. It was suggested therefore that the ideal candidate for this job should be highly educated, preferably having research experience, often combined with a background in IT or LIS. The participant acknowledged that most support activities related to data management require knowledge of the research processes, as far as becoming part of research teams, especially because only researchers know their data, even if they need help from data stewards (Kvale, 2021).

A recent event in this field, organized by the Research Data Management Forum that works under the auspices of the Digital Curation Centre (DCC), shed light on the potential scope of activities of stewards working in research institutions:

- "Delivering policy change on RDM, FAIR and open science across the organization,
- Providing support, training, and consultancy to enable change in research workflows and practices,
- Liaising between policy stakeholders and data centers, repositories, or other units of the research data infrastructure to deliver compliant services" (Research Data Management Forum, 2021).

Research data management education

The first focal point of education that we decided to choose is research data management (RDM). It concerns researchers, teaching staff members, students, and librarians.

When looking for ideas about objectives and curricula of education, we can take the example of a sociotechnical approach to education of data librarians that is part of the masters programs in LIS at City University of London. Focusing both on library applications and issues, and on finding meaningful information in data, Robinson and Bawden (2017) posited that the education of the library workforce requires appreciating technical and socioethical competences, the latter containing knowledge of legal issues, taking into account that many of these have been known well to librarians in the paper-based world. Accordingly, their course aimed to "provide the technical and philosophical background required to collect, store, describe, structure, manage and share information effectively in the digital society, by engaging with the deluge of digital data, and distilling information from it" (p. 316).

Southall and Scutt (2017, pp. 307–308) demonstrated that the basic learning objectives of a course for librarians evolved from recognizing RDM as vital to promoting it. They believe that "professional development in RDM leads to stronger integration with existing traditional library support without compartmentalizing queries." These issues have become "part of the everyday knowledge base in a similar way to how literature search advice was once a specialized area."

In order to build a baseline capacity, Wittenberg, Sackmann and Jaffe (2018, pp. 325–328) identified three learning outcomes for educating academic library staff. The first one is recognizing data management questions and responding confidently, followed by actively building and participating in an RDM community of practice. The third outcome aimed at understanding the data management needs of researchers in their domain and incorporating that awareness into reference and instruction work. All these learning outcomes served as a starting point for devising a framework of activities and specific outcomes, depending on skill level, need, and interest.

They had to make decisions, whether to focus on a discipline (some disciplines) or use generalized curriculum models, and found that in-person, domain-based RDM trainings were more successful than general ones.

Clare et al. (2019) provided 24 RDM case studies from research institutions worldwide. These case studies describe innovative activities to engage with data that could be implemented in other institutional settings as well.

The literature review of Tayler and Jafary (2021, pp. 83–88) focused on training-the-trainer models of the RDM training landscape. In such models, the targeted audience can be librarians, as well as researchers; thus the pedagogical models should be shaped accordingly. As they point out, "there are significant advantages to delivering discipline-specific or targeted RDM training. However, a generalized approach to RDM training may be favoured due to perceived scalability."

Actively building and participating in a research data management community of practice is the next task that needs to be supplemented by understanding the data management needs of researchers in their domain and incorporate that awareness into reference and instruction work.

The learning outcomes were set as follows:

1. Developing the ability of recognizing data management questions and responding confidently,
2. Actively building and participating in an RDM community of practice,
3. Understanding the data management needs of researchers in their domain and incorporating that awareness into reference and instruction work.

The process for instructional design for each workshop included, among others:

- Identifying and determining learning outcomes,
- Planning content delivery dates and session locations,
- Developing communication plans for the training,
- Contacting guest speakers (where applicable),
- Designating responsibility for curriculum and developing assessment questions,
- Reviewing feedback,
- Reviewing the training cycle.

Parallel to such decisions, there is a need to choose among online, in-person, or hybrid delivery. Online training modules are among the most desired ones among RDM professionals, because they are perceived to allow flexibility in accommodating working schedules. Moreover, the quality of online trainings is usually improved when paired with in-person instruction (Tang & Hu, 2019; Tayler & Jafary, 2021).

Pothier and Condon (2020, pp. 129–135) identified seven key fields of skills necessary for future business professionals. In our opinion, most of them are valid also beyond the business sphere.

1. Data organization and storage,
2. Understanding data, used in business contexts,
3. Evaluating the quality of data sources,
4. Interpreting data,
5. Data-driven decision-making,
6. Communicating and presenting effectively with data,
7. Data ethics and security.

These skills also correlate with data literacy, as it is "at home in the library and information science fields because it is complementary to information literacy. In fact, most of the attempts to characterize data literacy competencies have synthesized previous efforts to identify data literacy competencies."

Smits and Teperek (2020) studied how graduate students from different scientific backgrounds think about RDM. These students were aware of privacy issues associated with data processing, but showed confusion about the definition of RDM. They felt unanimously that good data management is important for research reproducibility. However, their approach to data management was ad hoc, i.e., when facing problems or challenges with their data, they dealt with them intuitively.

Munasinghe and Svirsky (2021) report on data analytics and data science courses for a diverse group of students from business, IT, engineering, computer science, physics, biochemistry, and earth science. The data science course taught students how to accomplish viable data-driven solutions based on authentic, real-world problems. The courses intended to simulate the real-world environment where professionals from various backgrounds come together to solve real-world problems.

An interesting lesson can be learned from the experiences of the master's programs of the University of Pisa (Italy). At this university the interdisciplinary nature of the educational environment has already been recognized; thus they developed a blend of computer science and traditional and digital humanities methods and data science education programs.

The academic subjects that contribute to LIS education are mainly part of DH and DS, with significant overlaps between them, but important contributions also come from traditional humanities and computer science. These courses are most frequently selected by digital humanities students, but they are also taken by students of data science and business informatics (Casarosa, Ruggieri, Salvatori, Simi, & Turbanti, 2020). This shows that there is apparently an intersection involving LIS, data science, and the digital humanities.

The Bodleian Libraries at the University of Oxford offer a model of how the integration of RDM training with other areas of open scholarship might be achieved. They promote RDM by using an established framework, called iSkills (https://libguides.bodleian.ox.ac.uk/workshops) that covers a wide range of topics, related information discovery, and scholarly communications, and is aimed at specific disciplines and cuts across them. This example clearly demonstrates not only the width and depth of undertakings, but shows the potential of collaborating with other organizations, or simply using open access materials from other institutions.

Following are a few examples of educational initiatives and materials that can be accessed through the iSkills system or the UK Data Service (https://www.ukdataservice.ac.uk):

- Versioning and authenticity control strategy,
- Legal and ethical issues,
- Storing research data,
- Making secondary use of statistics and data vs databases or data collections, which should be used for academic purposes, or shared with anyone else, destroyed once used, and not reused for other projects,
- Data collections acquired through individual negotiation, or orphaned following the completion of a project.

Education for data literacy

The second focal point of education is data literacy that requires giving increased attention to the needs of students, although it should take care of the needs of all academic target audiences.

"Datasets do not speak for themselves, because they require context for analysis and interpretation" (Gebre & Morales, 2020, p. 19). This statement becomes more obvious in the context of data quality (see Chapter 3), but it is also meaningful in the sense of the appropriateness of requiring relevant education. Without it, the practical use of datasets is limited, either due to the lack of data literacy skills, or because of the "data speaks for itself" mentality. Providing data literacy education is an important part of academic libraries' mission, even though it is addressed under different names.

Data-driven decision-making, mentioned in a previous subsection, merits some attention, because—among others—it has received special attention in business studies. This is similar in education, where there is an express interest in data-driven decision-making. This is emphasized by Mandinach and Gummer (2013, p. 32), who add that in educational environments, data literacy is often confused with assessment literacy, yet "assessments as data collected from classroom, school, district, and state measures of student achievement are only one type of data that enter into the decision-making process. Data literacy requires knowledge of other data, such as perception, motivation, process, and behaviour." The necessity of taking a wide range of sources in order to make appropriate decisions is also true beyond education or the business sphere.

One of the required abilities of data literacy is translating vast amounts of data into abstract concepts, as well as understanding databased reasoning. Context here is utterly important. Dissociating data from its context and the loss of context make reuse difficult or even impossible (Schneider, 2013).

Data literacy education is viable only if we focus on all related competencies. However, from among the numerous competencies, data citation requires particular attention because there is no overall view that would bring together diverse aspects of the education of professionals (Silvello, 2018). In this way, fostering critical understanding of data in higher education is a key issue. In cases that go beyond purely academic research aims, contributing to civic participation and to the promotion of social justice requires fostering a series of transversal skills through critically oriented pedagogies, which give attention to technical data skills, as well as information and media literacies in a broader content of citizenship. In order to achieve this, understanding and analyzing data from a range of sources, and reflecting on it, should be developed (Atenas, Havemann, & Timmermann, 2020). A set of data citation principles has been published by the Data Citation Synthesis Group (FORCE 11) (https://www.force11.org/group/data-citation-synthesis-working-group) and there is an exhaustive guide for data citation from the Digital Curation Centre (DCC) (Ball & Duke, 2015).

If we focus on a somewhat different, but also important, characteristic of data literacy education, i.e., the emphasis on the creation of data (Koltay, 2015), then we can accept the possibility of combining data literacy with the major elements of research data management. Such a program was devised by Carlson & Stowell Bracke, 2015. Instead of creating a syllabus at the onset of the program, they developed weekly lesson plans for their data literacy course, designed for graduate students, that contained the following final topics:

- Data management planning,
- Data lifecycle models,
- Discovery and acquisition,
- Description and metadata,

- Data security and storage,
- Data copyright and licensing,
- Data sharing and file-naming conventions,
- Data management and documentation,
- Data visualization,
- Data repositories,
- Data preservation, and
- Data publication and curation.

Read, Larson, Gillespie, Oh, and Surkis (2019) devised a series of online RDM education modules for librarians. The first module, called The Story of Data, aims at providing background information to foster a concrete understanding of the different forms that research data can take and data pathways from conceptualization to collection to processing to analysis. It involves, among others, distinguishing between different categories of data and identifying the full range of data products. The second module is an introduction to the research data lifecycle, mapping out the full range of data management activities, and their relationship to the research process. It pinpoints the data management needs at each stage of the data lifecycle. The authors also explain how to distinguish between data management needs for reproducibility purposes versus reuse, and identify different options for researchers to disseminate their data. The third module focuses on understanding researchers working in practices of bench science and clinical research. When addressing the climate of RDM, they underline that researchers should recognize requirements, incentives, and resources for managing and sharing of research data. Data documentation best practices require outlining all the components of a research workflow that should be documented, including best practices for file naming, documenting variable names in a data dictionary, as well as applying best practices to variable selection and naming. Collecting data and providing metadata for research data are also important. The module on storage, preservation, and sharing directs attention to selecting the appropriate storage solution(s) for datasets, communicating the difference between storage and preservation, and assessing the pros and cons for sharing different types of data.

A case study of Watts, Sare, and Hubbard (2020, p. 258) explored the benefits of data literacy education in a research lab setting at Texas A&M University (College Station, TX). They found that data literacy instruction provided by a sole data librarian involves a pitfall, related to "understanding and aligning the disciplinary context to learning experiences for multiple audiences." The root of this problem is in the impossibility of aligning the educational content with all discipline-based data standards. Disciplinary expertise may not be essential for one-shot, multidisciplinary workshops, but is a key component of applied settings such as a scientific laboratory.

Education for data literacy unavoidably involves addressing issues of the education for other literacies and related abilities. As already discussed to some extent in the previous chapter, the ACRL Framework for Information Literacy for Higher Education (ACRL, 2016) laid the foundations of an up-to-date approach not only to data literacy, but inspired the thinking on the ways of education for data literacy. Taking the closeness of these two literacies, this is a natural development, which resulted in the development of the preliminary core conceptual framework for data literacy by Abner (2020). Her data literacy framework is based on the idea of threshold concepts. As she put it, a person who understands a threshold concept "can ask precise questions and will exhibit certain behaviours that show a deep level of understanding" (p. 19).

There are a few ideas related to the ACRL Framework's nature, which are applicable to approaching data literacy through Abner's framework. According to Drabinski (2016, p. 383), the Framework is "not primarily about discovering something once and for all, but instead about opening up a terrain of exploration, spurring the construction of new knowledges that build on and extend what has come before." As Miller (2018, p. 412) puts it, the Framework "prompts librarians to take an important detour from the immediate practical application or transliteration attempts of the Frames" by requiring they visit a more theoretical landscape in their development and integration of a new pedagogical praxis.

As it is known, information literacy is often influenced by disciplinary indicators. However, "defining authority where basic indicators, such as credentials or type of publications, are insufficient or unclear can be challenging." The multidisciplinary nature of some fields may further complicate the task of assigning authority appropriately.

In addition to this, defining authority should not start with a discussion of specific criteria. Instead, "students should be given the opportunity to check their understanding of authority in general and to contrast it with authority as it is defined in the discipline or disciplines within which they operate" (Pullman, 2018, pp. 417–418).

The majority of competency frameworks for data literacy underline the importance of evaluation. For instance, Carlson, Fosmire, and Miller (2011, p. 633) acknowledged relatively early the importance of "understanding what data mean, including how to read graphs and charts appropriately, draw correct conclusions from data, and recognize when data are being used in misleading or inappropriate ways." This idea reminds us that—while data in varied workplace contexts may be used differently than in pure research and academic scholarship pursuits—understanding the sources, uses, and applications of data is a universal competency.

Data literacy education has varied foci. For instance, Calzada Prado and Marzal (2013) focused on applying evaluation criteria such as authorship, methods of collection, analysis, and interpretation. Ridsdale et al. (2015)

associated this competency with the ability to assess trustworthiness in regard to data sources. Duffner-Ylvestedt and Rayner (2016) directed our attention to the importance of educating data literate individuals in the context of their subject area. Students should identify relevant data for their field of study, understand how data is connected to the publication process, and be able to explain what open data is and how open data influences science. Applying critical thinking strategies to data and understanding the challenges of reusing data, as well as having awareness of the importance of metadata and citation, are crucial if we want to handle data appropriately. The correctness of this was proved in a study of job advertisements, where Khan and Du (2018) found that the skills of research assistance and critical thinking and problem solving appeared in more job advertisements than knowledge of data sources or a statistical software package.

A literature review by Ince, Hoadley, and Kirschner (2019) of recent works related to information literacy practices geared toward the needs of doctoral students shows that there are four key skill sets needed to effectively manage and publish their research: information literacy, information management, knowledge management, and knowing the lifecycle of scholarly communication. This is one of the reasons why it can be said that we should not separate different literacies from each other. Moreover, it is underlined that we should be cautious when separating the education for different literacies from each other. This is also true regarding the education for research data management. Neither should data literacy be separated from the education for other literacies. Regarding these ideas, we can take the example of media literacy, a broad conceptualization that is not necessarily media specific, because it defines primary skills necessary for social development, including critical thinking (Pfaff-Rüdiger & Riesmeyer, 2016). We should namely accept that the convergences between media literacy and information literacy gave enough incentive to properly recognizing the importance of avoiding silos without denying the existence differences between these two literacies. It is therefore important to clarify that data literacy does not replace digital literacies, although due to datafication and the reliance on it, would suggest this not so speak about the intensity of data-driven epistemologies that rely on comparison, pattern recognition, prediction, and analytics (Kitchin, 2014). Networked interactivity also dictates the need for more algorithmic thinking (Pangrazio & Sefton-Green, 2020), which is one of the components of computational thinking (mentioned earlier and described in Chapter 5 on literacies), is complementing problem solving with cooperativity, creativity, and critical thinking (Doleck, Bazelais, Lemay, Saxena, & Basnet, 2017).

Partially due to the growing importance of research data, the skills of evaluating visual information have become undoubtedly crucial. This is one of the reasons why data visualization has appeared in several parts of this book. Due to its significance and complexity, data visualization must be incorporated into information literacy and data literacy education. To evaluate a particular solution to data visualization, we have to inspect its quality, accuracy, and reliability, not forgetting the fact that "the power of the image will overwhelm the substantive content that it represents" (Womack, 2014, pp. 12–14). Visualization services are not new to academic libraries and can have many forms. However, data visualization and the related services should be distinguished from visualizing library data.

Visual literacy in general is not necessarily intuitive or inherent, thus the skills and abilities required for carrying out related activities need to be taught, because their acquisition is not automatic (ACRL, 2011). Accordingly, visual reading and writing skills include defining the need for finding images, then evaluating, interpreting, analyzing, and citing them. Images must be used effectively and ethically (Hattwig, Bussert, Medaille, & Burgess, 2013). These skills are divided into the thematic groups of visual communication, visual creation and image production, and image use (Kędra, 2018).

In many regards, visual literacy operates on a continuum between recognizing, understanding, applying, analyzing, evaluating, and creating. Nonetheless, its processes are different from verbal (textual) literacy, because it does not always begin with recognition and understanding. The reason for this is that images may be evaluated and interpreted before someone is able to name them. Moreover, evaluation can be successful even if recognition is missing (Emanuel & Challons-Lipton, 2013). A proper level of visual literacy is achieved "when an individual understands the various ways in which a discipline uses these components to represent a particular way of knowing. Owing to the complex nature of visual representations, the activities through which visual literacy is developed have high cognitive load." When engaging in disciplinary discourse, students and researchers are making use of various semiotic resources, such as equations, graphs, and diagrams, each of which provide access to a different facet of disciplinary ways of knowing. This requires the development of discursive fluency that allows making connections between representations and the various facets of disciplinary knowledge (Offerdahl, Arneson, & Byrne, 2017, pp. 1–3).

Similarities and convergences in the pedagogical approaches to the education for literacies

Although it may be precarious to separate educational approaches from the description of literacies, the next part of this chapter examines some approaches to teaching and learning literacies, in particular information literacy, first of all by indicating some of the circumstances that define the actual information (and) media environment. As indicated earlier, most of what we say here about information literacy applies to data literacy as well, thanks to the kinship of information literacy and data literacy, which is also true for most members of the literacy family. Nonetheless, in this subsection, we address only a selection of related issues without striving for an exhaustive treatment of them. The problems of education in academia are often perceived by recognizing the tension between subject knowledge and (otherwise often lacking) pedagogical skills.

There are at least two basic abilities that are especially relevant when engaging in education for varied literacies. The first one is stressing the need to make responsible choices when accessing, comprehending, and sharing information and data. The second, closely connected one is emphasizing the importance of analyzing messages by identifying their author(s), purpose, and point of view, and evaluating the quality and credibility of their content (Hobbs, 2010). Although these abilities are not new and come from media literacy education, they could have been taken from the practice of information literacy and fit perfectly to data literacy. This is obviously not by accident, but reflects the convergence that was underlined in Chapter 5 of this book, titled "New literacies in the age of convergence."

With this background, our starting point is that the development of varied literacies, including data literacy into education, would benefit from considering different views on the relationships between information literacy and learning (Onyancha, 2020). Moreover, we can accept that librarians and data providers share similar data literacy goals (Jefferson, Stierholz, Fontichiaro, & Hoelter, 2020). Nonetheless, this subsection aims only at indicating and very briefly describing the nature of a small set of approaches and methods for educational activities related to data and information literacies.

Social-constructionist approaches

According to Lloyd (2017, p. 92), IL can be seen as a set of skills and the abilities that foster successful engagement with codified forms of knowledge. In this space, IL can be viewed as an object and outcome of learning, both related to developing a curricula and effectiveness in teaching. This is one of the distinct narratives, which puts a strong emphasis on operationalization and application and attempts to quantify the practice. In this space, IL is shaped by teaching and learning within specific settings.

However, there is a conceptual space as well where "IL is conceived primarily from a socially oriented perspective, which emphasises the qualitative nature of information experiences in relation to specific contexts." In this perspective, the focus is on how IL shapes "people's ways of knowing and their ability to construct information landscapes."

All this is also true for data literacy, whereby at present data literacy is more often seen through the first instrumental and operational approach to literacies. Nonetheless (as said earlier) the conceptual framework for data literacy by Abner (2020) builds on the ACRL Framework for Information Literacy for Higher Education (ACRL, 2016). This means that we can count on the application of social-constructionist approaches.

Phenomenographic models, for instance, see learning as an activity of constructing meaning, without classifying learning experiences as right or wrong (Limberg, Sundin, & Talja, 2013). Such approaches focus on participants' experiences by asking questions to find out not only why something happens, but also how and what the participants do and how they feel about it (Morrison & Secker, 2017).

In the *sociocultural* perspective, IL is conceptualized in a collective view that considers situated learning and complex social realities (Lloyd, 2017). Communities of practice, i.e., groups of people sharing similar goals and interests, play a role here.

The *discourse analytic* perspective sees language as central to social life, providing thus a key to social functioning (Walton & Cleland, 2017). Both the sociocultural and the discourse analytic perspectives portray learning as a social activity that uses tools, practices, and conditions for meaning-making (Pilerot, 2016). Obviously, these literacies have closer ties to textual information than to numeric data, so their application may prove less efficient.

Inquiry-based learning is regarded as "student centred and a means of fostering critical thinking where learners construct knowledge for themselves as they attempt to make sense of their experiences"; it is an example of a constructivist approach (Walton & Cleland, 2017, pp. 583–584).

Inquiry-based learning is an "educational strategy in which students follow methods and practices similar to those of professional scientists in order to construct knowledge" (Keselman, 2003, p. 898). According to Pedaste et al. (2015), it is also a process of discovering new causal relations, with the learner formulating hypotheses and testing them by conducting experiments and/or making observations. Often it is viewed as an approach to solving problems and involves the application of several problem-solving skills. Based on these ideas, a learning framework that can improve the skills, related to "identifying problems, formulating questions and hypotheses, planning and carrying out experiments, collecting and analysing data, presenting the results, and drawing conclusions," was elaborated. The framework of inquiry-based learning contains the main phase of orientation, the goal of which is stimulating curiosity about a topic and addressing a learning challenge through a problem statement. It is followed by conceptualization, which is the process of stating theory-based questions and/or hypotheses. The investigation phase consists of planning, exploration or experimentation, collecting, and analyzing data. Comparing inferences, based on data, is the basis for drawing conclusions. Discussion is the process of presenting findings, and controlling the whole learning process or its phases by engaging in reflective activities.

Inquiry generally refers to a range of educational approaches, geared toward taking students' interests into account, "including students asking questions and undertaking authentic, discipline-based or cross-disciplinary extended units of individual or small group investigation, and expanded numbers and types of roles for learners and teachers" (Borovay, Shore, Caccese, Yang, & Hua, 2019, pp. 74–76).

Hoffman, Beatty, Feng, and Lee (2017) described an inquiry-based learning course for skills instruction. This course merits attention because it was also an example of making use of embedded librarianship, in-class instruction, formative and summative assessments, and a semiflipped teaching approach. In this case, being embedded meant that librarians not only provided instruction but also contributed to course design, content, pedagogy, and student assessment. The flipped classroom is way of preparing the learners before class (e.g., in the form of a short video), followed by in-class activities that reinforce the content as learned. Moreover, the course integrated library research skills with writing skills.

A cognitive approach

Cognitive models operate with two forms of representation: informational knowledge (knowing what) and operational knowledge (knowing how). They state that knowledge unfolds on the level of data, names, labels, simple statements, and then achieves a higher level of integration, i.e., the level of rules, theories, and formal systems.

Supporting the development of informational knowledge in education requires making learners aware of the importance of selecting information already before they engage in information processing. Present here are a purposefully organized learning landscape and a learning activity. The presence of cognitive economics is also needed, because the learner does not waste cognitive decision-making time on information processing if it is not useful for that particular learning activity. Both the learners and the teacher must differentiate between visual and propositional contents, and different learning situations. The learner should be able to distinguish between elementary information (statements) and complex information (principles) to decide what kind of information is needed to solve the given problem in a purposeful and economical way.

The development of informational knowledge would not be possible without the functioning of operational knowledge. The working of the operational knowledge, i.e., the performance of the representations, is determined by skills, as well as abilities of task solving and problem solving. Forming a system of informational knowledge within the cognitive process materializes in thinking operations: analysis, synthesis, enhancement, comparison, abstraction, and generalization.

The development of operational knowledge requires the teachers to form the learners' lower and higher psychic operations, also including analysis and synthesis, but supplemented by searching association, comparison, sequence recognition, identification, detecting relationships, as well as evaluation, application, restructuration, classification, and organization. Diverse paths of intellectual activities should be purposefully planned and practiced. There is also a need for raising consciousness of the learners' personal activities and self-reflection, because this is the only way to develop metacognition and conscious planning of information (Szőke-Milinte, 2020). Making use of thinking operations and thinking about the ways of developing operational knowledge, mentioned earlier, justify the application of such a model.

Connectivism

As Siemens (2004, pp. 1–5) explains it, cognitivism and constructivism are broad learning theories, often utilized in the creation of instructional environments. These theories, however, were developed in a time when learning was not impacted through technology.

He emphasizes that connectivism responds to this development, whereby many learners will move into a variety of different, possibly unrelated, fields over the course of their lifetime. As formal education no longer comprises the majority of our learning, we engage in informal learning, which becomes a significant aspect of our learning experience.

Learning occurs in a variety of ways: through communities of practice, personal networks, and completion of work-related tasks. It is also a continual lifelong process, in which learning and work are related or identical activities.

Technology is altering (rewiring) our brains and the tools that we use define and shape our thinking, and there is a link between individual and organizational learning. Many of the processes previously handled by learning theories (especially in cognitive information processing) can now be off-loaded to, or supported by, technology.

Siemens is undoubtedly correct when stating that "knowing how" is being supplemented by understanding "where to find" knowledge. If we accept this, we are going back to the roots of information retrieval and information literacy. When using Kuhlthau's Information Search Process (ISP) framework to illustrate ways that librarians can assist students and collaborate with data providers to improve the data search process, Jefferson et al. (2020) indirectly confirm that searching for information should be mapping this process across three domains: what researchers are doing (physical actions), thinking (cognition), and feeling (emotion) (p. 7).

Returning to the argument of Siemens, we can look at the central principles of connectivism:

- Learning is a process of connecting specialized nodes or information sources,
- Decision-making is itself a learning process,
- When specialized nodes and information sources are composed of digital, electronic, and online resources, learning and technology are inextricably linked,
- The ability to perceive connections between numerous perspectives, opinions, and concepts is central to learning,
- Developing and maintaining connections is necessary in order to facilitate continuous learning,
- Evaluating information prior to engaging with that information is a metaskill that is applied before learning begins. These evaluative decisions regarding the worthiness of learning specific information represent a learning process,
- Learning and knowledge rests in the diversity of opinions,
- Learning may reside in nonhuman appliances,
- Capacity to know more is more critical than what is currently known,
- Nurturing and maintaining connections is needed to facilitate continual learning,
- Ability to see connections between fields, ideas, and concepts is a core skill (pp. 5–6).

When explaining the use of digital information literacy badges, Raish and Rimland (2015, pp. 91–92) expressed the view that the basic assumptions of connectivism fit well with the tenets of information literacy. They also explain that digital badges "can serve both as a node in a learner's network to better connect them to social networks for possible connections and as a way to structure activities." Digital badges are scripts for students in their learning environment and as visual representations. They can be seen as microcredentials that are similar to badges, certifying skills at a more granular level than grades or transcripts, and—in other words—as representations of an accomplishment, interest, or affiliation.

The scholarship of teaching and learning (SoTL)

Teaching forms a major part of many librarians' professional identity, although they rarely have formal pedagogic training. Most of these librarians have never thought that they will be occupied with handling books or navigating databases. They are facing instead tasks related to knowledge under construction in people's heads. This is the reason why Coonan (2019, p. 128) says that the Scholarship of Teaching and Learning (SoTL) "offers teaching librarians not only the tools and insights required to conduct formal inquiry into their practice, but also the means to confront deeper questions about themselves and their identities and about the roles and responsibilities these roles bring." This is true, although we must see that, since its first use by Boyer (1990), the meaning of SoTL seems to remain elusive (Shawa, 2020).

SoTL is by definition a form of professional development. As Shulman (2006) put it:

> "The scholarship of teaching and learning invites faculty…to view teaching as serious, intellectual work, ask good questions about their students' learning, seek evidence in their classrooms that can be used to improve practice, and make this work public so that others can critique it, build on it, and contribute to the wider teaching commons" (p. ix).

This is one of the reasons why we do not have a canonical definition for SoTL, while its purposes are clear enough. SoTL is namely an applied form of scholarship that changes with the distinct styles of its practitioners. As Felten, MacMillan, and Ruelle (2019, p. 246) put it, professional development "should embody these criteria, but they are not a simple recipe for individual action or strategic programming. Instead, these sketch the outlines of what scholarly professional development might be, leaving plenty of room for variation to reflect the local contexts and the particular needs of individuals and groups involved in this work." SoTL-informed professional development reflects the essential criteria for any scholarly activity:

- Clear goals of the activities, as well as explicit and clear purposes,
- Adequate preparation, drawing on relevant scholarship and expertise,
- Appropriate methods, aligned with goals and context, and good practice in the field,
- Significant and meaningful results,
- Effective presentation of the purposes, methods, and results of the activity, shared with appropriate audiences, and
- Reflective critique by those who are involved in assessing the experiences and outcomes to inform future action.

Shulman (2012, p. 50) also underlined the pragmatic rationale for this form of inquiry: "by engaging in purposive reflection, documentation, assessment and analysis of teaching and learning, and doing so in a more public and accessible manner," teachers not only support the improvement of their own teaching, but also contribute to a community of scholarly teachers.

As a framework, SoTL looks not only at the content of education, i.e., what students are learning, but "the ways in which they learn, the processes through which knowledge is constructed, thresholds crossed, and complex understandings developed" (Coonan, 2019, p. 129).

The principles of good practice require SoTL to be inquiry-focused, grounded in context, methodologically sound, and appropriately public. Partnership with students may enable engaging in critical practices.

SoTL offers opportunities for the professional development of librarians. Focusing on student learning requires understanding "forms of evidence and deeper insights into what learning is, how it looks, and how all the attendant factors beyond what the teacher is doing may affect it." Context requires researchers to understand more deeply the environment in which someone teaches and the learners with whom they work, making learning in this situation uniquely transferable to or from the settings of other studies. Deeper engagement with familiar methods may require the exploration and a clear understanding of new methods. Making work appropriately public may "stretch skills in new directions and engage librarians with unfamiliar discourses" (Felten et al., 2019, p. 251).

Obviously, being reflective in order to enhance our practice within and across the separate and related domains of professional activity is required not only in SoTL programs, but can help us make "sense of and achieve coherence in the different areas of our individual careers and personal lives" (Corrall, 2017, p. 23).

MOOCs

Massive Open Online Courses (MOOCs) provide opportunities and resources for reaching distant learners without the constraints of their geographical locations and time. While the most helpful aspect of MOOCs are their hands-on nature and flexibility of the course schedules, the less intuitive nature of interfaces may affect the learning experience negatively (Liu, Kang, & McKelroy, 2015).

MOOCs seem to be providing a promising channel for developing IL skills of students. Co-funded by the European Union, a generic, multidisciplinary MOOC was developed. It is available online at https://informationliteracy.eu. This MOOC contains six modules:

1. Orienting in an information landscape,
2. Research is a journey of inquiries,
3. The power of search,
4. Critical information appraisal,
5. Information use: the right and fair way,
6. Let us create something new based on information and share it!

Using several standard models of information literacy, 16 core concepts to be addressed in the coursework have been identified:

- Understanding the information environment,
- Using digital tools effectively,
- Recognizing information needs,

- Being familiar with relevant information resources,
- Finding and accessing information,
- Critically evaluating information and information sources,
- Critically evaluating online interactions and online tools,
- Managing information,
- Collaborating in information handling,
- Sharing digital content ethically,
- Becoming independent and self-directed and lifelong learners,
- Learning to learn and developing metacognition,
- Understanding ethical issues of information,
- Presenting and communicating information,
- Creating information products,
- Synthesizing information and creating new knowledge (Dreisiebner et al., 2020, p. 782).

Although the authors of this MOOC do not mention data literacy, we believe that its closeness to IL gives enough ground to think that our claim, characterizing our opinion throughout this book, is applicable: teaching materials for varied literacies can be tailored to the need for data literacy education.

A piece of circumstantial evidence for this, presented by the existence of a recent, free online RDM training material, not called a MOOC but similar to such courses, offered by the Research Data Management Librarian Academy (RDMLA, https://www.canvas.net/browse/simmonsu/courses/research-data-management), includes the following units:

- Foundations of RDM,
- Navigating research culture,
- Advocating for RDM in libraries,
- Launching data services,
- Project management and assessment,
- Data analysis and visualization tools,
- Coding tools,
- Platform tools,
- Delivering data management training,
- Data copyright, licensing, and privacy,
- Data curation and archiving.

This small selection of methods and programs offers insight into the width and depth of possibilities for educational activities, be they offered by data librarians or by other professionals, targeting RDM or data literacy.

Breaking out of silos

Although avoiding silos is not a method but a problem to be solved, it seems to be appropriate to put it at the end of our subsection on educational approaches.

Reviewing information and media literacy side by side reveals opportunities for integration that will allow professionals to address the problems outside of the silo mentality that still often exists between these two concepts and practices (Leaning, 2019).

An interesting example of looking for an opportunity to step outside silos is establishing faculty learning communities (FLCs). They are a particular kind of communities of practice, meant to foster the transformation of higher education institutions into learning organizations. An FLC has been established at the Medical Library, Oakland University William Beaumont School of Medicine (Rochester, MI), among others, to:

- Create a sense of community for teaching staff members across disciplines with teaching, continuous learning, and educational research needs,
- Provide a safe, social space for them to interact with one another and to engage in collective and collaborative learning, as well as sharing their experience and concerns (Mi, 2015).

A new graduate certificate course in Petroleum Data Management (PDM) also shows the problem of avoiding silos. The developers of this course tried to cover the gap in education and training of data, rooted in the fact that the participants are usually trained much more on the use of algorithms and technical tools than on data governance and ethical concerns. Part of the challenge was that PDM is an interdisciplinary field combining geoscience, computing, information management, and data management. In practice, this means that, while PDM recruits staff from these fields, many staff members will subsequently return to their original, clearly defined career path. PDM is "not an established discipline, or part of one, and cannot offer a fulfilling career path on a par with established disciplines, such as engineering." Thus the challenge of developing this course was to deliver it while taking into account both the necessary academic and "real-world" rigor, while ensuring that critical thinking and reflection were embedded in the learning outcomes of the course (Cousins, Reid, & Tait, 2019, p. 1214).

References

Abner, K. (2020). Data literacy as digital humanities literacy: Exploration of threshold concepts. In M. Brooks, M. Hubbard, J. Perkins, & J. Russell (Eds.), *Literacies in a digital humanities context: A dh+ lib special issue* (pp. 19–22). dh+ lib.

Abraham, R., Schneider, J., & Vom Brocke, J. (2019). Data governance: A conceptual framework, structured review, and research agenda. *International Journal of Information Management*, *49*, 424–438. https://doi.org/10.1016/j.ijinfomgt.2019.07.008.

ACRL. (2011). *ACRL visual literacy competency standards for higher education*. Chicago, IL: Association of College and Research Libraries. Retrieved from http://www.ala.org/acrl/standards/visualliteracy.

ACRL. (2016). *Framework for information literacy for higher education*. Chicago, IL: Association of College and Research Libraries. Retrieved from http://www.ala.org/acrl/standards/ilframework.

Atenas, J., Havemann, L., & Timmermann, C. (2020). Critical literacies for a datafied society: Academic development and curriculum design in higher education. *Research in Learning Technology, 28*. https://doi.org/10.25304/rlt.v28.2468.

Azeroual, O. (2020). Treatment of bad big data in research data management (RDM) systems. *Big Data and Cognitive Computing, 4*(4), 29. https://doi.org/10.3390/bdcc4040029.

Azeroual, O., & Lewoniewski, W. (2020). How to inspect and measure data quality about scientific publications: Use case of Wikipedia and CRIS databases. *Algorithms, 13*(5), 107. https://doi.org/10.3390/a13050107.

Ball, A., & Duke, M. (2015). How to cite datasets and link to publications. In *DCC how-to guides*. Edinburgh: Digital Curation Centre. Retrieved from https://www.dcc.ac.uk/guidance/how-guides.

Baškarada, S., & Koronios, A. (2017). Unicorn data scientist: The rarest of breeds. *Program: Electronic Library and Information Systems, 51*(1), 65–74. https://doi.org/10.1108/PROG-07-2016-0053.

Berthold, M. R. (2019). What does it take to be a successful data scientist? *Harvard Data Science Review, 1*(2). https://doi.org/10.1162/99608f92.e0eaabfc.

Borgman, C. L. (2018). Open data, grey data, and stewardship: Universities at the privacy frontier. *Berkeley Technology Law Journal, 33*, 365–412. https://doi.org/10.15779/Z38B56D489.

Borovay, L. A., Shore, B. M., Caccese, C., Yang, E., & Hua, O. (2019). Flow, achievement level, and inquiry-based learning. *Journal of Advanced Academics, 30*(1), 74–106. https://doi.org/10.1177/1932202X18809659.

Boyer, E. L. (1990). *Scholarship reconsidered: Priorities of the professoriate*. Lawrenceville, NJ: Princeton University Press.

Burton, M., Lyon, L., Erdmann, C., & Tijerina, B. (2018). *Shifting to data savvy: The future of data science in libraries*. Pittsburgh, PA: University of Pittsburgh. Retrieved from https://d-scholarship.pitt.edu/33891/.

Calzada Prado, J. C., & Marzal, M.Á. (2013). Incorporating data literacy into information literacy programs: Core competencies and contents. *Libri, 63*(2), 123–134. https://doi.org/10.1515/libri-2013-0010.

Carlson, J., Fosmire, M., & Miller, C. C. (2011). Determining data information literacy needs: A study of students and research faculty. *Portal: Libraries and the Academy, 11*(2), 629–657. https://doi.org/10.1353/pla.2011.0022.

Carlson, J., & Stowell Bracke, M. (2015). Planting the seeds for data literacy: Lessons learned from a student-centered education program. *International Journal of Digital Curation, 10*(1), 95–110. https://doi.org/10.2218/ijdc.v10i1.348.

Casarosa, V., Ruggieri, S., Salvatori, E., Simi, M., & Turbanti, S. (2020). Educational ecosystems for Information Science: The case of the University of Pisa. *Education for Information, 36*(2), 119–138. https://doi.org/10.3233/EFI-190330.

Clare, C., Cruz, M., Papadopoulou, E., Savage, J., Teperek, M., Wang, Y., … Yeomans, J. (2019). *Engaging researchers with data management: The cookbook (epub)*. Open Book Publishers. https://doi.org/10.11647/OBP.0185.

Coonan, E. M. (2019). Inside/outside/in between: Librarians and SoTL research. In *The grounded instruction librarian: Participating in the scholarship of teaching and learning* (pp. 128–137). Chicago, IL: Association of College and Research Libraries.

Corrall, S. (2017). Crossing the threshold: Reflective practice in information literacy development. *Journal of Information Literacy, 11*(1), 23–53. https://doi.org/10.11645/11.1.2241.

Cousins, F., Reid, P., & Tait, E. (2019). Bridging the divide: Reflections on university-industry collaboration for the development of the graduate certificate in petroleum data management. *Journal of Documentation, 75*(6), 1213–1229. https://doi.org/10.1108/JD-12-2018-0206.

Cox, J. (2018). Communicating new library roles to enable digital scholarship: A review article. *New Review of Academic Librarianship*, *24*(3/4), 217–241. https://doi.org/10.1080/13614533.2018.1466342.

Davenport, T., & Patil, D. (2012). Data scientist: The sexiest job of the 21st century. *Harvard Business Review*, *90*(5), 70–76.

DCC. (2013). *Checklist for a data management plan. v.4.0*. Edinburgh: Digital Curation Centre. Retrieved from http://www.dcc.ac.uk/resources/data-management-plans.

Doleck, T., Bazelais, P., Lemay, D. J., Saxena, A., & Basnet, R. B. (2017). Algorithmic thinking, cooperativity, creativity, critical thinking, and problem solving: Exploring the relationship between computational thinking skills and academic performance. *Journal of Computers in Education*, *4*(4), 355–369. https://doi.org/10.1007/s40692-017-0090-9.

Drabinski, E. (2016). Turning inward: Reading the framework through the six frames. *College & Research Libraries News*, *77*(8), 382–384. https://doi.org/10.5860/crln.77.8.9537.

Dreisiebner, S., Polzer, A. K., Robinson, L., Libbrecht, P., Boté-Vericad, J. J., Urbano, C., … Pehar, F. (2020). Facilitation of information literacy through a multilingual MOOC considering cultural aspects. *Journal of Documentation*, *77*(3), 777–797. https://doi.org/10.1108/JD-06-2020-0099.

DSA. (2013). *Data science code of professional conduct*. Data Science Association. Retrieved from http://www.datascienceassn.org/code-of-conduct.html.

Duffner-Ylvestedt, N., & Rayner, J. (2016). Hooking up data with literacy: Creating an educational framework for Uppsala University Library. *Nordic Journal of Information Literacy in Higher Education*, *8*(1), 38–44. https://doi.org/10.15845/noril.v8i1.261.

Eclevia, M. R., Fredeluces, J. C. L., Maestro, R. S., & Eclevia, C. L., Jr. (2019). What makes a data librarian? An analysis of job descriptions and specifications for data librarian. *Qualitative & Quantitative Methods in Libraries*, *8*(3), 273–290. https://doi.org/10.5860/crl13-435.

Emanuel, R., & Challons-Lipton, S. (2013). Visual literacy and the digital native: Another look. *Journal of Visual Literacy*, *32*(1), 7–26. https://doi.org/10.1080/23796529.2013.11674703.

Federer, L. (2018). Defining data librarianship: A survey of competencies, skills, and training. *Journal of the Medical Library Association*, *106*(3), 294–303. https://doi.org/10.5195/jmla.2018.306.

Federer, L., Clarke, S. C., Zaringhalam, M., & Huerta, M. (2020). *Developing the librarian workforce for Data Science and Open Science*. Bethesda, MD: National Library of Medicine. Retrieved from https://osf.io/uycax/.

Felten, P., MacMillan, M., & Ruelle, J. (2019). SoTL difference: The value of incorporating SoTL into librarian professional development. In M. Mallon, L. Hays, C. Bradley, R. Huisman, & J. Belanger (Eds.), *The grounded instruction librarian: Participating in the scholarship of teaching and learning* (pp. 245–256). Chicago, IL: American Library Association.

Franklin, M. J., & Nicolae, D. L. (2021). A perspective on designing undergraduate education in the new discipline of Data Science. *Harvard Data Science Review*, *3*(2), 2. https://doi.org/10.1162/99608f92.8d7bff79.

Gebre, E. H., & Morales, E. (2020). How "accessible" is open data? Analysis of context-related information and users' comments in open datasets. *Information and Learning Sciences*, *121*(1/2), 19–36. https://doi.org/10.1108/ILS-08-2019-0086.

Gitelman, L. (2013). Introduction. In L. Gitelman (Ed.), *Raw data' is an oxymoron* (pp. 7–14). Cambridge, MA: MIT Press.

Gowen, E., & Meier, J. J. (2020). Research Data Management services and strategic planning in libraries today: A longitudinal study. *Journal of Librarianship and Scholarly Communication*, *8*(General Issue), eP2336. https://doi.org/10.7710/2162-3309.2336.

Hattwig, D., Bussert, K., Medaille, A., & Burgess, J. (2013). Visual literacy standards in higher education: New opportunities for libraries and student learning. *Portal: Libraries and the Academy*, *13*(1), 61–89. https://doi.org/10.1353/pla.2013.0008.

Henderson, M. (2020). Why you need soft and non-technical skills for successful data librarianship. *Journal of eScience Librarianship*, *9*(1), 2. e1183 https://doi.org/10.7191/jeslib.2020.1183.

Hobbs, R. (2010). *Digital and media literacy: A plan of action*. A White Paper on the Digital and Media Literacy Recommendations of the Knight Commission on the Information Needs of Communities in a Democracy Washington, DC: The Aspen Institute.

Hoffman, N., Beatty, S., Feng, P., & Lee, J. (2017). Teaching research skills through embedded librarianship. *Reference Services Review*, *45*(2), 211–226. https://doi.org/10.1108/RSR-07-2016-0045.

Ince, S., Hoadley, C., & Kirschner, P. A. (2019). The role of libraries in teaching doctoral students to become information-literate researchers. *Information and Learning Sciences*, *120*(3/4), 158–172. https://doi.org/10.1108/ILS-07-2018-0058.

Jefferson, C. O., Stierholz, K., Fontichiaro, K., & Hoelter, L. (2020). Considering data literacy using Kuhlthau's Information Search Process: Implications for librarians and data providers. *Journal of Business & Finance Librarianship*, *25*(3/4), 197–229. https://doi.org/10.1080/08963568.2020.1847556.

Johnston, L. R., Carlson, J., Hswe, P., Hudson-Vitale, C., Imker, H., Kozlowski, W., … Stewart, C. (2017). Data curation network: How do we compare? A snapshot of six academic library institutions' data repository and curation services. *Journal of eScience Librarianship*, *6*(1), e1102. https://doi.org/10.7191/jeslib.2017.1102.

Kędra, J. (2018). What does it mean to be visually literate? Examination of visual literacy definitions in a context of higher education. *Journal of Visual Literacy*, *37*(2), 67–84. https://doi.org/10.1080/1051144X.2018.1492234.

Kennan, M. A. (2016). Data Management: Knowledge and skills required in research, scientific and technical organisations. In *Paper presented at IFLA WLIC 2016—Columbus, OH—Connections. Collaboration. Community in Session 221—Science and Technology*. Retrieved from http://library.ifla.org/1466/1/221-kennan-en.pdf.

Kennan, M. A. (2017). 'In the eye of the beholder': Knowledge and skills requirements for data professionals. In *Vol. 22(4). Proceedings of RAILS—Research applications, information and library studies, 2016, School of Information Management, Victoria University of Wellington, New Zealand, 6-8 December (2016). Information research*. paper rails1601 http://InformationR.net/ir/22-4/rails/rails1601.html.

Keselman, A. (2003). Supporting inquiry learning by promoting normative understanding of multivariable causality. *Journal of Research in Science Teaching*, *40*(9), 898–921. https://doi.org/10.1002/tea.10115.

Khan, H. (2020). Exploring Data Science learning objectives in LIS education. In *ALISE 2020 juried papers* (pp. 317–325). Retrieved from https://www.ideals.illinois.edu/handle/2142/108828.

Khan, H. R., & Du, Y. (2018). *What is a data librarian? A content analysis of job advertisements for data librarians in the United States academic libraries*. ILA WLIC, Session 139. Retrieved from http://library.ifla.org/2255/1/139-khan-en.pdf.

Kitchin, R. (2014). Big Data, new epistemologies and paradigm shifts. *Big Data & Society*, *1*(1), 2053951714528481. 10.1177%2F2053951714528481.

Koltay, T. (2015). Data literacy: In search of a name and identity. *Journal of Documentation*, *71*(2), 401–415. https://doi.org/10.1108/JD-02-2014-0026.

Koltay, T. (2019). Accepted and emerging roles of academic libraries in supporting research 2.0. *Journal of Academic Librarianship*, *45*(2), 75–80. https://doi.org/10.1016/j.acalib.2019.01.001.

Kvale, L. H. (2021). Using personas to visualize the need for data stewardship. *College & Research Libraries*, *82*(3), 332–351. https://doi.org/10.5860/crl.82.3.332.

Leaning, M. (2019). An approach to digital literacy through the integration of media and information literacy. *Media and Communication*, *7*(2), 4–13. https://doi.org/10.17645/mac.v7i2.1931.

Limberg, L., Sundin, O., & Talja, S. (2013). Three theoretical perspectives on information literacy. *Human IT: Journal for Information Technology Studies as a Human Science, 11*(2), 93–130.

Liu, M., Kang, J., & McKelroy, E. (2015). Examining learners' perspective of taking a MOOC: Reasons, excitement, and perception of usefulness. *Educational Media International, 52*(2), 129–146. https://doi.org/10.1080/09523987.2015.1053289.

Lloyd, A. (2017). Information literacy and literacies of information: A mid-range theory and model. *Journal of Information Literacy, 11*(1), 91–105. https://doi.org/10.11645/11.1.2185.

Madrid, M. M. (2013). A study of digital curator competences: A survey of experts. *International Information and Library Review, 45*(3/4), 149–156. https://doi.org/10.1016/j.iilr.2013.09.001.

Mandinach, E. B., & Gummer, E. S. (2013). A systemic view of implementing data literacy in educator preparation. *Educational Researcher, 42*(1), 30–37. https://doi.org/10.3102/0013189X12459803.

Mi, M. (2015). Expanding librarian roles through a librarian initiated and facilitated faculty learning community. *Journal of Library Administration, 55*(1), 24–40. https://doi.org/10.1080/01930826.2014.978683.

Miller, S. D. (2018). Diving deep: Reflective questions for identifying tacit disciplinary information literacy knowledge practices, dispositions, and values through the ACRL Framework for information literacy. *Journal of Academic Librarianship, 44*(3), 412–418. https://doi.org/10.1016/j.acalib.2018.02.014.

Morrison, C., & Secker, J. (2017). Understanding librarians' experiences of copyright: Findings from a phenomenographic study of UK information professionals. *Library Management, 38*(6/7), 354–368. https://doi.org/10.1108/LM-01-2017-0011.

Munasinghe, T., & Svirsky, A. (2021). Engaging students in data literacy: Lessons learned from data intensive classrooms. In *13th ACM web science conference 2021 (WebSci '21 companion), June 21–25, 2021, virtual event, United Kingdom*. New York, NY: ACM. https://doi.org/10.1145/3462741.3466665.

Offerdahl, E. G., Arneson, J. B., & Byrne, N. (2017). Lighten the load: Scaffolding visual literacy in biochemistry and molecular biology. *CBE—Life Sciences Education, 16*(1), es1. https://doi.org/10.1187/cbe.16-06-0193.

Ogier, A. L., & Stamper, M. J. (2018). Data visualization as a library service: Embedding visualization services in the library research lifecycle. *Journal of eScience Librarianship, 7*(1), e1126. https://doi.org/10.7191/jeslib.2018.1126.

Oliver, J. C., Kollen, C., Hickson, B., & Rios, F. (2019). Data science support at the academic library. *Journal of Library Administration, 59*(3), 241–257. https://doi.org/10.1080/01930826.2019.1583015.

Onyancha, O. B. (2020). Knowledge visualization and mapping of information literacy, 1975–2018. *IFLA Journal, 46*(2), 107–123. https://doi.org/10.1177/0340035220906536.

Pangrazio, L., & Sefton-Green, J. (2020). The social utility of "data literacy". *Learning, Media and Technology, 45*(2), 208–220. https://doi.org/10.1080/17439884.2020.1707223.

Pedaste, M., Mäeots, M., Siiman, L. A., De Jong, T., Van Riesen, S. A., Kamp, E. T., … Tsourlidaki, E. (2015). Phases of inquiry-based learning: Definitions and the inquiry cycle. *Educational Research Review, 14*, 47–61. https://doi.org/10.1016/j.edurev.2015.02.003.

Pfaff-Rüdiger, S., & Riesmeyer, C. (2016). Moved into action. Media literacy as social process. *Journal of Children and Media, 10*(2), 164–172. https://doi.org/10.1080/17482798.2015.1127838.

Pilerot, O. (2016). Connections between research and practice in the information literacy narrative: A mapping of the literature and some propositions. *Journal of Librarianship and Information Science, 48*(4), 313–321. https://doi.org/10.1177/0961000614559140.

Poole, A. H. (2017). "A greatly unexplored area": Digital curation and innovation in digital humanities. *Journal of the Association for Information Science and Technology, 68*(7), 1772–1781. https://doi.org/10.1002/asi.23743.

Pothier, G. W., & Condon, P. B. (2020). Towards data literacy competencies: Business students, workforce needs, and the role of the librarian. *Journal of Business & Finance Librarianship, 25*(3/4), 123–146. https://doi.org/10.1080/08963568.2019.1680189.

Pullman, E. (2018). Applying the framework as a reflective tool: A teacher learner perspective. *College & Research Libraries News, 79*(8), 416–419. https://doi.org/10.5860/crln.79.8.416.

Raish, V., & Rimland, E. (2015). Employer perceptions of critical information literacy skills and digital badges. *College & Research Libraries, 77*(1), 87–113. https://doi.org/10.5860/crl.77.1.87.

Rawlings-Goss, R. (2019). *Data Science careers, training, and hiring. A comprehensive guide to the data ecosystem: How to build a successful Data Science career program, or unit*. Cham: Springer International. https://doi.org/10.1007/978-3-030-22407-3_1.

Read, K. B., Larson, C., Gillespie, C., Oh, S. Y., & Surkis, A. (2019). A two-tiered curriculum to improve data management practices for researchers. *PLoS One, 14*(5), e0215509. https://doi.org/10.1371/journal.pone.0215509.

Research Data Management Forum. (2021). *Digital Curation Centre*. Retrieved from https://www.dcc.ac.uk/events/rdmf.

Rice, R., & Southall, J. (2016). *The data librarian's handbook*. London: Facet.

Ridsdale, C., Rothwell, J., Smit, M., Ali-Hassan, H., Bliemel, M., & Irvine, D. (2015). *Strategies and best practices for data literacy education. Knowledge synthesis report*. Halifax, NS: Dalhousie University.

Robinson, L. (2016). Between the deluge and the dark age: Perspectives on data curation. Editorial *Alexandria, 26*(2), 73–76. https://doi.org/10.1177/0955749016661067.

Robinson, L., & Bawden, D. (2017). "The story of data": A socio-technical approach to education for the data librarian role in the CityLIS library school at City, University of London. *Library Management, 38*(6/7), 312–322. https://doi.org/10.1108/LM-01-2017-0009.

Sabharwal, A. (2017). Digital humanities and the emerging framework for digital curation. *College & Undergraduate Libraries, 24*(2–4), 238–256. https://doi.org/10.1080/10691316.2017.1336953.

Schneider, R. (2013). Research data literacy. In *Worldwide commonalities and challenges in information literacy research and practice* (pp. 134–140). Cham: Springer International.

Scroggins, M. J., Pasquetto, I. V., Geiger, R. S., Boscoe, B. M., Darch, P. T., Cabasse-Mazel, C., … Borgman, C. L. (2019). Thorny problems in data (-intensive) science. *Communications of the ACM, 63*(8). https://doi.org/10.1145/3408047.

Semeler, A. R., & Pinto, A. L. (2020). Data librarianship as a field. *Transinformação, 32*, e200034. https://doi.org/10.1590/2318-0889202032e200034.

Shawa, L. B. (2020). Advancing the scholarship of teaching and learning using learning theories and reflectivity. *CEPS Journal, 10*(1), 191–208. https://doi.org/10.26529/cepsj.298.

Shulman, L. (2006). Forward. In T. Hatch (Ed.), *Into the classroom: Developing the scholarship of teaching and learning* (pp. vii–x). San Francisco, CA: Jossey-Bass.

Shulman, L. (2012). From Minsk to Pinsk: Why a scholarship of teaching and learning? *Journal of the Scholarship of Teaching and Learning, 1*(1), 48–53. Retrieved from https://scholarworks.iu.edu/journals/index.php/josotl/article/view/1582.

Siemens, G. (2004). Connectivism: A learning theory for the digital age. *International Journal of Instructional Technology and Distance Learning, 2*(1), 3–10. Retrieved from http://www.itdl.org/Journal/Jan_05/article01.htm.

Silvello, G. (2018). Theory and practice of data citation. *Journal of the Association for Information Science and Technology, 69*, 6–20. https://doi.org/10.1002/asi.23917.

Smits, D. A. B., & Teperek, M. (2020). Research Data Management for master's students: From awareness to action. *Data Science Journal, 19*(1). https://doi.org/10.5334/dsj-2020-030.

Song, I. Y., & Zhu, Y. (2017). Big data and data science: Opportunities and challenges of iSchools. *Journal of Data and Information Science, 2*(3), 1–18. https://doi.org/10.1515/jdis-2017-0011.

Southall, J., & Scutt, C. (2017). Training for Research Data Management at the Bodleian Libraries: National contexts and local implementation for researchers and librarians. *New Review of Academic Librarianship, 23*(2–3), 303–322. https://doi.org/10.1080/13614533.2017.1318766.

Szőke-Milinte, E. (2020). *Információ—Média(tudatosság)—Műveltség. A Z generáció tanulása.* Budapest: Pázmány Péter Katolikus Egyetem (in Hungarian).

Tang, R., & Hu, Z. (2019). Providing Research Data Management (RDM) services in libraries: Preparedness, roles, challenges, and training for RDM practice. *Data and Information Management, 3*(2), 84–101. https://doi.org/10.2478/dim-2019-0009.

Tayler, F., & Jafary, M. (2021). Shifting Horizons: A literature review of Research Data Management train-the-trainer models for library and campus-wide research support staff in Canadian institutions. *Evidence Based Library and Information Practice, 16*(1). https://doi.org/10.20381/cmc6-6y81.

Thomas, C. V., & Urban, R. J. (2018). What do data librarians think of the MLIS? Professionals' perceptions of knowledge transfer, trends, and challenges. *College & Research Libraries, 79*(3), 401–423. https://doi.org/10.5860/crl.79.3.401.

Trepanier, C., Shiri, A., & Samek, T. (2019). An examination of IFLA and Data Science Association ethical codes. *IFLA Journal, 45*(4), 289–301. https://doi.org/10.1177/0340035219849614.

Virkus, S., & Garoufallou, E. (2019). Data science from a library and information science perspective. *Data Technologies and Applications, 53*(4), 422–441. https://doi.org/10.1108/DTA-05-2019-0076.

Virkus, S., & Garoufallou, E. (2020). Data science and its relationship to library and information science: A content analysis. *Data Technologies and Applications, 54*(5), 643–663. https://doi.org/10.1108/DTA-07-2020-0167.

Walton, G., & Cleland, J. (2017). Information Literacy: Empowerment or reproduction in practice? A discourse analysis approach. *Journal of Documentation, 73*(4), 582–594. https://doi.org/10.1108/JD-04-2015-0048.

Wang, L. (2018). Twinning data science with information science in schools of library and information science. *Journal of Documentation, 74*(6), 1243–1257. https://doi.org/10.1108/JD-02-2018-0036.

Watts, J., Sare, L., & Hubbard, D. E. (2020). Collaborative data literacy education for research labs: A case study at a large research university. *Collaborative Librarianship, 12*(3). Retrieved from https://digitalcommons.du.edu/collaborativelibrarianship/vol12/iss3/.

Wildgaard, L., Vlachos, E., Nondal, L., Larsen, A. V., & Svendsen, M. (2020). *National coordination of data steward education in Denmark: Final report to the National Forum for Research Data Management (DM Forum).* Danish e-Infrastructure Cooperation (DeiC). https://doi.org/10.5281/zenodo.3609516.

Wittenberg, J., Sackmann, A., & Jaffe, R. (2018). Situating expertise in practice: Domain-based data management training for liaison librarians. *Journal of Academic Librarianship, 44*(3), 323–329. https://doi.org/10.1016/j.acalib.2018.04.004.

Womack, R. (2014). Data visualization and information literacy. *IASSIST Quarterly, 38*(1), 12–17. https://doi.org/10.7282/T3X92CZF.

Index

Note: Page numbers followed by *f* indicate figures.

9780128244753